Multivariate scaling methods and the reconstruction of social spaces

Alice Barth
Felix Leßke
Rebekka Atakan
Manuela Schmidt
Yvonne Scheit (eds.)

Multivariate scaling methods and the reconstruction of social spaces

Papers in honor of Jörg Blasius

Verlag Barbara Budrich
Opladen • Berlin • Toronto 2023

This work was supported by the Open Access Publication Fund of the University of Bonn.

© 2023 This work is licensed under the Creative Commons Attribution 4.0 (CC-BY 4.0). It permits use, duplication, adaptation, distribution and reproduction in any medium or format, as long as you give appropriate credit to the original author(s) and the source, provide a link to the Creative Commons license and indicate if changes were made. To view a copy of this license, visit https://creativecommons.org/licenses/by/4.0/

© 2023 Dieses Werk ist bei der Verlag Barbara Budrich GmbH erschienen und steht unter der Creative Commons Lizenz Attribution 4.0 International (CC BY 4.0): https://creativecommons.org/licenses/by/4.0/. Diese Lizenz erlaubt die Verbreitung, Speicherung, Vervielfältigung und Bearbeitung unter Angabe der UrheberInnen, Rechte, Änderungen und verwendeten Lizenz.

The use of third party material in this book does not indicate that it is also subject to the Creative Commons licence mentioned. If the material used is not subject to the aforementioned Creative Commons licence and the action in question is not permitted by law, the consent of the respective rights holder must be obtained for further use. Trademarks, company names, generally descriptive terms etc. used in this work may not be used freely. The rights of the respective rights holder must be observed and use is subject to the rules of trademark law, even without separate reference.

This book is available as a free download from www.budrich.eu (https://doi.org/10.3224/84742764).

© 2023   by Verlag Barbara Budrich GmbH, Opladen, Berlin & Toronto
         www.budrich.eu
         ISBN   978-3-8474-2764-3
         eISBN 978-3-8474-1856-6
         DOI   10.3224/84742764

Die Deutsche Nationalbibliothek – CIP-Einheitsaufnahme
Ein Titeldatensatz für die Publikation ist bei der Deutschen Nationalbibliothek erhältlich: https://portal.dnb.de.

Verlag Barbara Budrich GmbH
Stauffenbergstr. 7. D-51379 Leverkusen Opladen, Germany

86 Delma Drive. Toronto, ON M8W 4P6 Canada
www.budrich.eu

Cover design by Bettina Lehfeldt, Kleinmachnow, Germany – www.lehfeldtgraphic.de
Typesetting by Michaela Moreels, Dillingen, Germany
Printed in Europe on FSC®-certified paper by docupoint GmbH, Barleben, Germany

Table of contents

Introduction 7
Alice Barth

Part I: Statistical properties of correspondence analysis and related methods 13

Michael Greenacre: P(CA) 13
Jörg Breitung: Dr. Strangelove or how I learned to stop worrying
and love correspondence analysis 29
Frédéric Lebaron, Brigitte Le Roux, Aykiz Dogan:
Applying Combinatorial Inference in GDA. The Case of European
Central Bank Governing Council Members (1999-2022) 46

Part II: Sociological applications of correspondence analysis and other multivariate scaling methods 69

Alice Barth, Rebekka Atakan, Felix Leßke, Yvonne Scheit,
Manuela Schmidt: Reconstructing a scientist's "space of social relations"
via multiple correspondence analysis 69
Martin Fritz & Yasemin El-Menouar: Religiosity and climate
change beliefs in Europe 83
Rahim Hajji, Simone Pollak, Gunnar Voß, Ulrike Scorna, Jessica Schäfer:
Health inequality, working conditions in dual vocational training
and educational inequality – An analysis using categorical principal
components analysis and hierarchical cluster analysis 106

Part III: Social inequality and change in (urban) space 131

Jens S. Dangschat: Gentrification as a self-producing and self-reenforcing process on the macro, meso and micro level 131

Karl M. van Meter: CARME – The rise and fall of a sociological space to the east 149

Nina Baur, Elmar Kulke: Social milieus in urban space 164

Part IV: Data quality and statistical education 195

Rainer Diaz-Bone: Sociological perspectives on teaching statistics in sociology – disciplinary complexities, interdisciplinary challenges and data worlds 195

Pieter C. Schoonees, Patrick J. F. Groenen, Michel van de Velden, Hester van Herk: Effects of Visual Priming on Rating Scale Usage 213

Part V: Reminiscences, anecdotes and greetings 227

Miriam Trübner, Andreas Mühlichen:
Jörg Blasius – An Academic Life in Numbers 227

Clemens Albrecht: Structure: follow culture!
Eine kurze Soziologie der akademischen Freundschaft 230

Nina Baur: Persönliche Notiz 232

Jens S. Dangschat: Zwei wichtige Erlebnisse – für Jörg und für mich 234

Robert Helmrich: An Jörg 235

Rahim Hajji: An Jörg 237

Yasemin El-Menouar: Für Jörg 238

Martin Fritz: An Jörg 239

Patrick J. F. Groenen: Travels with Jörg 240

Michael Greenacre: My 39 years with Jörg 241

Wendelin Strubelt: Eine Annäherung an dich, Jörg, an uns Reisende 245

Author information 249

Index 256

Introduction
Alice Barth

This edited volume is dedicated to Jörg Blasius in honor of his retirement from the chair of sociology and social research methods at the University of Bonn. Throughout his career, Jörg has combined statistical methods, most prominently methods for the analysis of categorical data, with sociological questions on social inequality, lifestyles, gentrification processes, and urban neighborhoods; he has also focused on the assessment of the quality of survey data and applied a broad range of social research methods. This plurality of research interests is also reflected in the book at hand, where long-term collaborators (who are usually long-term friends at the same time), former PhD students, and research assistants have written papers on research topics they share with Jörg.

In total, the contributions show that Jörg does not only look back on an extremely active and productive scientific career, but he also has an extraordinary record in supporting fellow scientists and fostering international networks. Editing a volume that represents the full scope of Jörg's scientific network would have been a quite impossible task – with this book, we have aimed at drawing a representative sample of the huge population of friends, colleagues and collaborators. The book is organized into five parts, each of which reflects one of Jörg's areas of interest.

The first part of the book treats the discussion of *statistical properties of correspondence analysis and related methods.* Jörg began exploring correspondence analysis as a scaling method for multivariate categorical data in his student days (see Blasius 1987 and *Michael Greenacre*'s reminiscence on page 241); the publication of his habilitation thesis (Blasius 2001) cemented his status as Germany's leading scholar in this area. Michael Greenacre's contribution elaborates on the relationships between principal component analysis and correspondence analysis as methods of dimension reduction and the visualization of data tables. Michael Greenacre and Jörg Blasius have shared a passion for correspondence analysis as a means for analyzing categorical data for more than 30 years, resulting, among others, in the organization of 9 conferences (since 2003 under the name of CARME – Correspondence Analysis and Related MEthods), four books (Greenacre & Blasius 1994, 2006; Blasius & Greenacre 1998, 2014) and two special issues (Blasius et al. 2009; Balbi et al. 2017; see also p. 244). Complementary to Michael's contributions as a long-term correspondence analysis specialist, *Jörg Breitung* approaches correspondence analysis from the perspective of an "alien from the planet Econometrics". He also compares principal component analysis and correspondence analysis, focusing then on the properties of the

7

chi-square distance measure and suggesting possible alternative measures of distance. Finally, *Frédéric Lebaron, Brigitte LeRoux* and *Aykiz Dogan* discuss possibilities for drawing combinatorial inferences in the framework of geometric data analysis. They exemplify their approach by constructing a "social space" of members of the European Central Bank's Governing Council.

Their contribution also segues into the next part of the book, where *sociological applications of correspondence analysis and other multivariate scaling methods* are discussed. Alongside to his methodological and statistical interest in the analysis of categorical data, Jörg Blasius has applied (multiple) correspondence analysis to a myriad of empirical examples from sociology and related disciplines. Among these are the social distribution of young people's perception of work (Thiessen & Blasius 2002), social norms in poor neighborhoods (Friedrichs & Blasius 2003), a Swiss health space (Lengen & Blasius 2007), the effects of cultural distance, free trade agreements, and protectionism on perceived export barriers (Korneliussen & Blasius 2008), response effects of item phrasing (Blasius & Friedrichs 2009), and the (strong) relationship between perceived corruption, trust and, interviewer behavior (Blasius & Thiessen 2021a), to name just a few. Many of his works are theoretically and methodologically grounded in the oeuvre of Pierre Bourdieu, who used correspondence analysis to objectify social phenomena and relationally (re)construct social spaces and fields. Jörg has often applied the "social space approach", assessing relations between lifestyles and economic and cultural capital (e.g. Blasius & Winkler 1989; Blasius 1994; Blasius 2000; Blasius & Friedrichs 2008; Blasius & Mühlichen 2010). The empirical investigation of social space has also been the topic of two conferences (Cologne 1998 and Bonn 2015) that brought together scientists who work empirically with Bourdieusian concepts from all over Europe (Blasius & Schmitz 2017; Blasius et al. 2019). Following this methodological approach, *Alice Barth, Felix Leßke, Rebekka Atakan, Manuela Schmidt* and *Yvonne Scheit* have reconstructed Jörg's "space of social relations" by collecting data from more than one hundred of Jörg's friends, colleagues, and collaborators and analyzing these using multiple correspondence analysis. The resulting space is characterized by the contrast between a public/professional and a private/leisure pole and the frequency of respondents' contact with Jörg. In a different vein, *Martin Fritz* and *Yasemin El-Menouar* discuss the relationship between attitudes towards climate change and religiosity, which they empirically assess by means of multiple correspondence analysis using data from the European Social Survey. *Rahim Hajji, Simone Pollak, Gunnar Voss, Ulrike Scorna* and *Jessica Schäfer* analyze connections between health inequality, working conditions in vocational education, and inequality of education by applying categorical principal component analysis and hierarchical cluster analysis. Apart from presenting sociological applications of multivariate scaling methods for categorical data, all three contributions in this section are authored by former or current PhD students of Jörg Blasius, demonstrating his success in "spreading the correspondence analysis virus".

The third part of the book addresses another topic that has accompanied Jörg throughout his academic life: the analysis of *social inequality and change in (urban) space*. Starting with his PhD thesis (Blasius 1993), he has frequently worked on the subjects of gentrification and social inequality in urban neighborhoods (e.g., Friedrichs & Blasius 2001; Blasius & Friedrichs 2007; Friedrichs & Blasius 2016; Blasius et al. 2016; Friedrichs & Blasius 2020), including several DFG-funded projects on life in disadvantaged neighborhoods, neighborhood effects, and gentrification processes. The majority of these projects and publications were conducted together with Jürgen Friedrichs, who awakened Jörg's interest for urban sociology in his student days, and was one of his most valued collaborators and friends until his sudden death in 2019. *Jens Dangschat*, who has also known Jörg since his days as a student assistant at the University of Hamburg (see page 234), reflects on developments, debates, and results in gentrification research and suggests, inspired by Bourdieusian concepts, a "macro-micro-meso model of place and space" to organize and instruct analyses of gentrification. A historical view on a very peculiar space is presented in *Karl van Meter's* contribution: the rise and fall of a sociological space in the east. From the perspective of an eyewitness, he portrays how the Soviet Union and East Germany came to host two international sociological conferences shortly before the dissolution of the Eastern bloc, narrates the experiences he, Jörg, and other colleagues made there, and describes the potential political controversy of a correspondence analysis which was to be carried out on Soviet leaders' power networks. *Nina Baur* and *Elmar Kulke* present a comparative analysis of housing and eating practices and the structure of food retail in the three cities Nairobi, Singapore, and Berlin. They discuss how class-specific practices of distinction vary in the different social contexts, and how they interact with characteristics of the urban space. Next to their shared interest in urban sociology, Nina and Jörg are close collaborators who together have accomplished the mammoth task of editing the German "Handbook of Empirical Social Research Methods" (*Handbuch Methoden der empirischen Sozialforschung*; Baur & Blasius 2022): Currently in its third edition, it has grown to two volumes comprising 122 chapters by 161 authors over 1743 pages – while the print version weighs more than 3 kg and is thus handy both as a reference work and a weapon, the online version (of the first edition) has accumulated the unbelievable figure of more than 22 million downloads (see also Nina's personal note on page 232). Untiringly, Jörg and Nina are already preparing a fourth, even larger edition.

Accordingly, the fourth section of the book features contributions that dovetail with Jörg's interest in imparting knowledge on *data quality and statistical education*. When explaining the significance of statistics, Jörg attaches importance to embedding statistical numbers in a sociological context (Blasius & Thiessen 2021b). In line with this, *Rainer Diaz-Bone* elaborates on teaching statistics in sociology. In his contribution, he discusses disciplinary complexities, interdisciplinary challenges, and the perspectives provided by differentiating between different "data worlds". Jörg has repeatedly shown how the conditions of producing survey data (study architecture, institutional practices, and respondent behavior) are reflected in differential quality of datasets and indicators that are widely used in the social sciences. In particular, he worked on this theme with his Canadian friend Victor Thiessen (Blasius & Thiessen 2001a, 2001b, 2006, 2012, 2021a), who regrettably passed away in 2016. Unsurprisingly, Jörg applies correspondence analysis and related scaling methods when testing data quality, as do *Pieter C. Schoonees, Patrick J.F. Groenen, Michel van de Velden* and *Hester van Herk* in assessing how the priming of respondents with the task of estimating the capacity of a cylinder affects the subsequent use of rating scales.

Finally, there is a fifth section which does not present research works, but rather *reminiscences, anecdotes and greetings* from some of Jörg's long-standing companions, reflecting the prominent role friendships and cordial personal relations have always played for him. *Miriam Trübner* and *Andreas Mühlichen* – as former research assistants cognizant of Jörg's fondness for numbers and the visualization of data – present figures for several aspects of his working life, such as his lifetime total number of lectures or the most frequently used words in his publications. *Clemens Albrecht* reflects on the role of personal friendships in academia. *Jens Dangschat, Nina Baur, Robert Helmrich, Rahim Hajji, Yasemin El-Menouar, Martin Fritz, Patrick Groenen* and *Michael Greenacre* recount their first encounters and memorable experiences with Jörg, and the volume is rounded off with a poem which *Wendelin Strubelt* has dedicated to him.

On behalf of the editorial team, I would like to thank all contributors, and especially the reviewers, for their efforts. We also thank the University of Bonn for funding the Open Access publication of this volume.

References

Balbi, Simona/Blasius, Jörg/Greenacre, Michael (eds.) (2017): Correspondence Analysis and Related Methods. Special issue of „Italian Journal of Applied Statistics", 29, 2–3.

Baur, Nina/Blasius, Jörg (eds.) (2022): Handbuch Methoden der empirischen Sozialforschung. Springer-Verlag.

Blasius, Jörg (1987): Korrespondenzanalyse – ein multivariates Verfahren zur Analyse qualitativer Daten. In: Historische Sozialforschung – Historical Social Research, 42/43, pp. 172–189.

Blasius, Jörg/Winkler, Joachim (1989): Gibt es die „feinen Unterschiede"? Eine empirische Überprüfung der Bourdieuschen Theorie. In: Kölner Zeitschrift für Soziologie und Sozialpsychologie, 41, pp. 72–94.

Blasius, Jörg (1993): Gentrification und Lebensstile. Eine empirische Untersuchung. Wiesbaden: Deutscher Universitäts-Verlag.

Blasius, Jörg (1994): Gentrification und lokale Lebensstile in Köln. Eine Anwendung der Korrespondenzanalyse. In: Allgemeines Statistisches Archiv, 78, pp. 96–113.

Blasius, Jörg/Greenacre, Michael (eds.) (1998): Visualization of Categorical Data. San Diego: Academic Press.

Blasius, Jörg (2000): Die Analyse von Lebensstilen mit Hilfe der Korrespondenzanalyse. In: Österreichische Zeitschrift für Soziologie, 25, 4, pp. 63–92.

Blasius, Jörg (2001): Korrespondenzanalyse. München: Oldenbourg.

Blasius, Jörg/Thiessen, Victor (2001a): Methodological Artifacts in Measures of Political Efficacy and Trust: A Multiple Correspondence Analysis. In: Political Analysis, 9, pp. 1–20.

Blasius, Jörg/Thiessen, Victor (2001b): The Use of Neutral Responses in Survey Questions. An Application of Multiple Correspondence Analysis. In: Journal of Official Statistics, 17, pp. 351–367.

Blasius, Jörg/Thiessen, Victor (2006): Assessing Data Quality and Construct Comparability in Cross-National Surveys. In: European Sociological Review, 22, pp. 229–242.

Blasius, Jörg/Friedrichs, Jürgen (2007): Internal Heterogeneity of a Deprived Urban Area and its Impact on Residents. In: Housing Studies, 22, pp. 753–780.

Blasius, Jörg/Friedrichs, Jürgen (2008): Lifestyles in distressed neighborhoods: A test of Bourdieu's "taste of necessity" hypothesis. In: Poetics, 36.1, pp. 24–44.

Blasius, Jörg/Friedrichs, Jürgen (2009): The effect of phrasing scale items in low-brow or high-brow language on responses. In: International Journal of Public Opinion Research, 21.2, pp. 235–247.

Blasius, Jörg/Greenacre, Michael/Groenen, Patrick/van de Velden, Michel (eds.) (2009): Correspondence Analysis and Related Methods. Special issue of „Computational Statistics and Data Analysis", 53.

Blasius, Jörg/Mühlichen, Andreas (2010): Identifying audience segments applying the "social space" approach. In: Poetics, 38.1, pp. 69–89.

Blasius, Jörg/Thiessen, Victor (2012): Assessing the Quality of Survey Data. London: Sage.

Blasius, Jörg/Greenacre, Michael (eds.) (2014): Visualization and Verbalization of Data. Boca Raton, Florida: Chapman & Hall.

Blasius, Jörg/Friedrichs, Jürgen/Rühl, Heiko (2016): Gentrifier and Pioneers in the Process of Gentrification. In: International Journal of Housing Policy, 16, pp. 50–69.

Blasius, Jörg/Schmitz, Andreas (2017): Conference Report – Empirical Investigation of Social Space II University of Bonn, 12–14 October 2015. In: Bulletin of Sociological Methodology/Bulletin de Méthodologie Sociologique, 133.1, pp. 65–70.

Blasius, Jörg/Lebaron, Frédéric/Le Roux, Brigitte/Schmitz, Andreas (eds.) (2019): Investigations of Social Space. Amsterdam: Springer International.

Blasius, Jörg/Thiessen, Victor (2021a): Perceived corruption, trust, and interviewer behavior in 26 European Countries. In: Sociological Methods & Research, 50.2, pp. 740–777.

Blasius, Jörg/Thiessen, Victor (2021b): Argumentieren mit Statistik: eine Einführung für das sozialwissenschaftliche Studium. UTB.

Friedrichs, Jürgen/Blasius, Jörg (2001): The Socio-Spatial Integration of Turks in Two Cologne Residential Neighbourhoods. In: German Journal of Urban Studies, 1.

Friedrichs, Jürgen/Blasius, Jörg (2003): Social Norms in Poverty Neighborhoods – Testing the Wilson Hypothesis. In: Housing Studies, 18, pp. 807–826.

Friedrichs, Jürgen/Blasius, Jörg (eds.) (2016): Gentrifizierung in Köln. Opladen: Barbara Budrich Verlag.

Friedrichs, Jürgen/Blasius, Jörg (2020): Neighborhood Change – Results from a Dwelling Panel. In: Housing Studies, 35, pp. 1723–1741

Greenacre, Michael/Blasius, Jörg (eds.) (1994): Correspondence Analysis in the Social Sciences. Recent Developments and Applications. London: Academic Press.

Greenacre, Michael/Blasius, Jörg (eds.) (2006): Multiple Correspondence Analysis and Related Methods. Boca Raton, Florida: Chapman & Hall.

Korneliussen, Tor/Blasius, Jörg (2008): The Effects of Cultural Distance, Free Trade Agreements, and Protectionism on Perceived Export Barriers. In: Journal of Global Marketing, 21 (3), pp. 183–195.

Lengen, Charis/Blasius, Jörg (2007): Constructing a Swiss Health Space Model of Self-perceived Health. In: Social Science & Medicine, 65, pp. 80–94.

Thiessen, Victor/Blasius, Jörg. (2002): The Social Distribution of Youth's Images of Work. In: Canadian Review of Sociology and Anthropology, 39.1, pp. 49–78.

Part I: Statistical properties of correspondence analysis and related methods

(P)CA

Michael Greenacre

Abstract

Principal component analysis and correspondence analysis are both methods of dimension reduction and visualization of data tables. Although they apply to different types of data, there are cases where they have similar results and sometimes even identical results. This chapter explores some relationships between the two methods.

1. Introduction

Correspondence analysis (CA) and principal component analysis (PCA) are closely related methods for visualizing a table of data. CA applies to categorical data in the form of frequencies, or a set of responses to nominal variables, but also to any non-negative ratio-scale data (see, for example, Greenacre 2016). PCA applies to interval-scale data, where computing means and variances makes sense (see the recent review by Greenacre et al. 2022). Since the logarithmic transformation converts ratio-scale observations to interval-scale ones, the PCA of a matrix of log-transformed data would be an alternative to performing CA, as long as the data were strictly positive. There would, however, be some noticeable differences in the results, since CA completely removes the size effect from the data, whereas PCA retains it, on a log-scale (Greenacre 2017). In other situations, a CA approach can substitute a PCA one, for example, when continuous data are coded into fuzzy variables and this transformed data set is analysed by CA rather than PCA to capture nonlinear patterns (Aşan and Greenacre 2011).

This article is concerned with the cases where PCA and CA would have a very close connection in their results and how this closeness can be explained.

The questions to be answered are the following: When is a CA the same as a PCA, and when is it almost the same?

2. When is a CA exactly a PCA?

CA weights the rows and columns proportionally to the row and column sums, whereas PCA does not, so if the table has constant row and column sums then one of the differences between them is eliminated. Such an example exists in the 4×4 table of positive integers in Fig. 1, the cryptogram sculpted next to the western entrance of the Sagrada Familia church in Barcelona, by the Catalan sculptor Josep Maria Subirachs. Every sum of four regular numbers (row sums, column sums, diagonals, 2×2 blocks in the corners and at the centre) is equal to 33, Christ's age at death. Hence, the CA row and column weights will all be ¼, chi-square distances will be Euclidean distances, as are the distances in PCA. The only difference between the CA and the PCA will then be a scaling factor, as seen in Fig. 2.

Figure 1: The cryptogram of Subirachs, sculpted on the wall of the Sagrada Familia in Barcelona. All "regular" sums of four numbers are equal to 33, in particular the row and column sums.

Source: Xavier via Wikimedia Commons, Public Domain

Figure 2: The CA and PCA of the cryptogram in Fig. 1

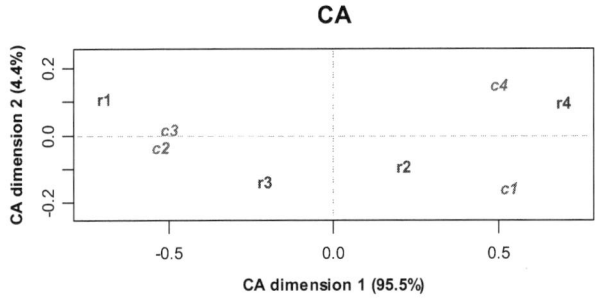

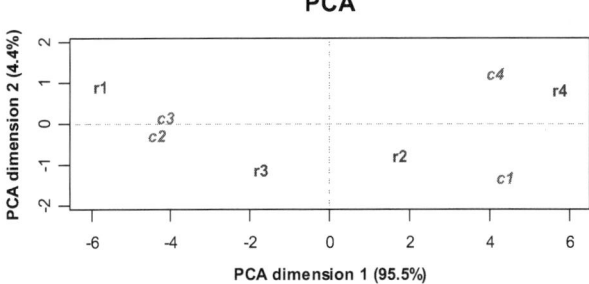

This simple example raises a number of interesting questions:

1. We know that the row points in both methods will be centred at the origin, since CA is double-centred and the row points in PCA are centred due to the automatic centring by column means. But how is it that the column points in the PCA (i.e., the usual variables) are also centred at the origin, as in CA?
2. PCA should have four dimensions, CA three. So what is the PCA's fourth dimension?
3. What is the scale factor that links the two results?

The answer to (1) is simple if one realizes that if you add a constant, say k, to the whole table, then all those regular sums, in particular the row and columns sums, will be simply increased by $4k$. For example, adding 1 to the whole table would make all the regular sums equal to 37. The usual PCA centring of the values in each column implies subtracting 33/4 from each element of the table, which means that the regular sums become 33−4(33/4) = 0. Hence, not only are the column sums zero, but also the row sums, i.e. the matrix is double-centred and the columns also have mean zero at the origin.

The answer to (2) flows directly from that of (1). Since the matrix is double-centred, the fourth dimension of PCA has a variance of zero – the PCA is also three-dimensional here.

The answer to (3) is exactly 33/4 = 8.25, for this particular form of the PCA computed in my *easyCODA* package (Greenacre 2018) in R (R Core Development Team 2022). Numerically, the singular values of the CA are 0.5154, 0.1103 and 0.0220, while those of the PCA are 4.2517, 0.9099, 0.0182 (and the fourth one of 0), with a factor difference of 8.25. If a chi-square distance is computed between two rows of the table, x and y, this would be between their row profiles $x/33$ and $y/33$, and with the expected values ¼ in the denominator (the values of the average profile). Hence, the square of the chi-square distance would be:

$$\sum_j (x_j/33 - y_j/33)^2 / (1/4) = \sum_j \frac{1}{4}(x_j - y_j)^2 / (33/4)^2 \quad (1)$$

On the right is the square of the Euclidean distance divided by 8.25^2, which means that the CA coordinates are the same as the PCA ones divided by 8.25. Note that the presence of the factor ¼ on the right hand side of (1) is a property of my personal version of PCA, implemented in easyCODA, which weights the rows and the columns by weights that sum to 1 (called *masses* in CA). These weights, which apply to the squared terms, would all be a constant ¼ in this equally weighted case, for the rows and the columns of this 4 × 4 table.

The equivalence of the two methods applies equivalently to any so-called "magic square" when the rows and columns have the same totals.

3. The weights in CA and PCA and the dimensionality threshold

The weights in CA are an important property of the method. In the usual implementations of PCA there are no weights on the rows and the columns, but it is customary to decompose the total variance in PCA, which inherently involves a weight of $1/(I-1)$ on the I rows in the variance calculation. The $I-1$ in the denominator is for unbiased variance estimation and is preferably ignored for our purpose of comparing PCA with CA, where we will simply use an equal weight of $1/I$ for each of the I rows. CA would have equal weights if all the rows had constant sums, which occurs if a compositional data matrix is analysed by CA, proportions adding up to 1 or percentages adding up to 100.

As for the columns, like the example in the previous section with 4 columns where the weights were said to be ¼, we will prefer PCA to have weights of $1/J$ for the J columns. This makes the definition of PCA more similar to that of CA, which has column weights adding up to 1. This also means that the column weights could be varied, as long as they are nonnegative and sum to 1. For standardized data with variables having variances of 1, the usual PCA definition involves a total variance of J. Over the J dimensions of a PCA, the average explained variance is thus 1, and there is a rule of thumb that the "interesting" dimensions correspond to explained variances (the PCA eigenvalues) greater than this average of 1. If the variables are weighted by $1/J$, the total variance, now the average variance, is equal to 1 and the eigenvalues will have an average of $1/J$. So the rule of thumb is then to choose those greater than 1 divided by the number of variables.

The above has a nice consequence when it comes to PCA's equivalent for categorical data, namely multiple correspondence analysis, or MCA. The original definition of MCA is the CA of an indicator matrix Z, the matrix of zeros and ones that indicate the responses of a sample of respondents to Q questions. The total number of columns of Z is J, that is, the total number of categories for all the questions is J. It is known that the total variance (often called inertia in the context of MCA and CA) is equal to the fixed amount $(J-Q)/Q$, irrespective of the data. Since it is also known that there are exactly $J-Q$ dimensions in the MCA, the average is $(J-Q)/Q$ divided by $(J-Q)$, equal to $1/Q$. So this agrees very nicely with the weighted version of PCA: in PCA the average is $1/J$, where J is the number of interval variables, and in MCA the average is $1/Q$, where Q is the number of categorical variables.

The average variance per dimension of $1/Q$ in MCA appears in three other contexts as well. The first is the definition of Cronbach's alpha reliability measure in the case of categorical data – see, for example, Greenacre (2016: pp. 159–160). If λ_k is the eigenvalue of the indicator form of MCA on the k-th axis, Cronbach's alpha is equal to:

$$\alpha = \frac{Q}{Q-1}\left(1 - \frac{1}{Q\lambda_k}\right) \qquad (2)$$

This means that it is only for $\lambda_k > 1/Q$ that the reliability is positive.

The second appearance of this $1/Q$ threshold is in adjusted MCA where the dimensions with eigenvalues above $1/Q$ can be adjusted to be consistent with least-squares approximation of the MCA solution with that of the off-diagonal crosstables of the Burt matrix. The formula is given by Greenacre (2016: equation (19.6)), where it should be pointed out that the notation λ_k is used there for the eigenvalue of the Burt matrix, which is the square of that of the indicator matrix as used here in (2).

Finally, the third appearance of $1/Q$ is in the conjecture, so far not proven or disproven, given in the last paragraph of Greenacre (2016: Epilogue), that the true dimensionality of multivariate categorical data is the number of eigenvalues (of the indicator matrix) greater than $1/Q$. This definition of dimensionality is the number of dimensions of the joint correspondence analysis (JCA) solution required to reproduce exactly all the pairwise two-way crosstables in the data (Greenacre 1988, 2016). JCA is different from MCA, in that its objective is specifically to approximate these pairwise crosstables, not to approximate the indicator or Burt matrix.

4. Doubled CA, CatPCA and MCA

For multivariate ordinal categorical data, for example responses to Q questions on a 5-point ordinal response scale, there are three options in CA, two at opposite extremes and one in-between. All options, however, aim to explain the maximum variance possible, subject to the constraints imposed by the analysis. The one extreme is to regard the scale points as nominal categories, and then perform an MCA – in this analysis each category is placed in any position in the solution (usually, two-dimensional), subject to the constraints that each set of five categories has the same weighted average of zero. At the other extreme is to treat the scale points as ordinal and equidistant, to perform a data doubling which codes the endpoints of the scale and then to perform CA, called a doubled CA. This forces the five scale points to be on a straight line and at equal intervals on this line, which crosses the origin at the variable's arithmetic average. In-between these two extreme methods, nonlinear PCA (de Leeuw: 2014), also called categorical PCA (CatPCA) presents a compromise. In CatPCA the scale points for each variable still lie on a straight line, but with the relaxation that the intermediate scale points can find their own positions between the endpoints, even at coinciding positions, but still in their ordinal order.

These three alternatives are demonstrated with a reduced data set extracted from the ISSP survey of the German population in 2020, on the topic of Environment (ISSP Research Group 2020). Only 120 respondents are chosen at random as an illustration, answering six questions about attitudes to the present environmental issues. Respondents with no missing responses were selected to avoid the separate issue of missing data. The six questions, in the form of statements are as follows:

A   Modern science will solve our environmental problems with little change to our way of life
B   We worry too much about the future of the environment and not enough about prices and jobs today
C   Almost everything we do in modern life harms the environment
D   People worry too much about human progress harming the environment
E   In order to protect the environment Germany needs economic growth
F   Economic growth always harms the environment

Response categories are 1=strongly agree, 2=agree, 3=neither agree nor disagree, 4=disagree, 5=strongly disagree.

The MCA in Figure 3 shows the typical arch that is often found in data with a strong gradient, in this case from pessimistic attitudes about the world's environment to optimistic ones. To follow the gradient, one has to move along the curve. At the pessimistic end there is strong disagreement to A, B, D, E, and agreement to F and C. Two actual responses are indicated at either end of the arch. Although the arch in MCA seems strange and one might think it is better straightened out, the one advantage is that points falling within the arch are a combination of the extreme positions. A point with response pattern [2 5 2 5 4 5] is shown, and it can indeed be seen that this respondent disagrees with both questions E and F, whereas those at the two ends take opposing views on these two statements.

Figure 3: MCA of the environmental survey data. The responses of three respondents are indicated next to their positions

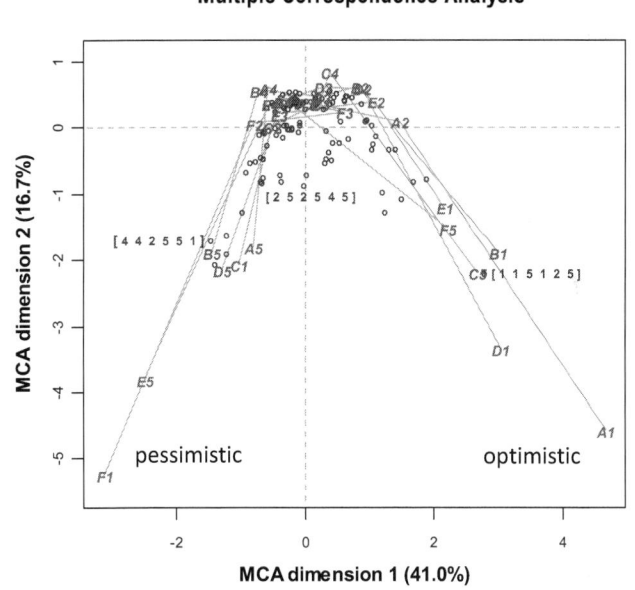

Figure 4: Doubled CA of the environmental survey data. The responses of a "pessimistic" and an "optimistic" respondent are indicated next to their positions.

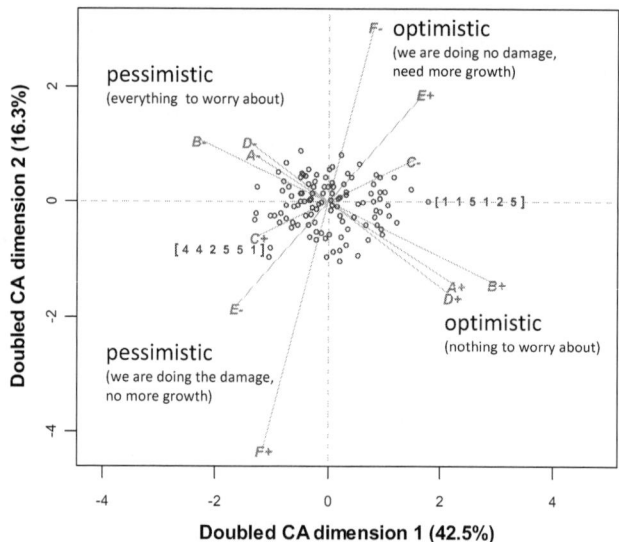

Figure 4 shows the doubled CA result. This classifies the statements into two groups: A, B and D, and C, E and F. The former group is about being concerned (or not) about what we are doing to the environment, and the latter about whether (or not) we are damaging the environment, especially through economic growth. The positions of the two extreme respondents in Figure 3 are also shown here.

Figure 5 shows the categorical PCA, using the function princals in the Gifi package (Mair and de Leeuw 2022) in R. Here, on the "optimistic" side we see a slightly different split of the statements compared to Figure 4, with respondents forming two "streams" to upper and lower right. The extreme optimist in the previous figures lies at the extreme of the upper right group, and is indeed the respondent the furthest to the right. An extreme point at lower right shows two "middle" (neither/nor) responses for the statements with strong disagreements in upper right response. Several other respondents in the lower right stream have these middle responses.

Figure 5: CatPCA of the environmental survey data. The responses of one "pessimistic" and two "optimistic" respondents are indicated next to their positions

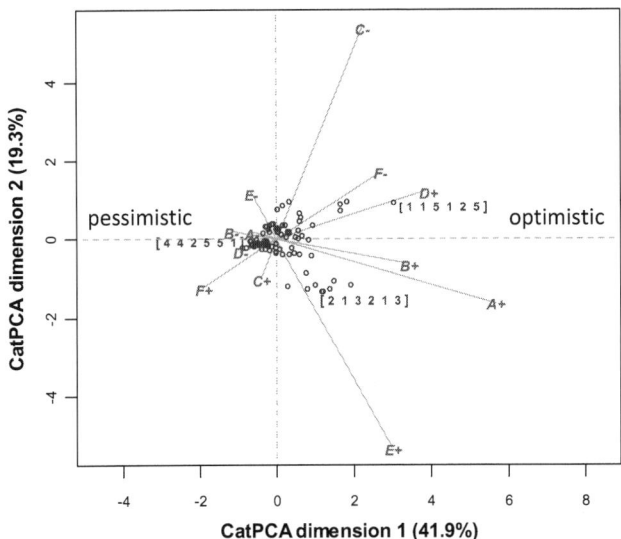

CatPCA has straightline displays of the different statements, but the category points are not equidistant – their positions are indicated in Figure 6, which displays the statement label at the agree side (i.e., category 1) and then the remaining category positions 2, 3, 4, and 5, in their order. In a few cases the categories coincide; for example, for statement C the categories 1, 2 and 3 are at the same position, which indicates that the method would prefer a different order, but the ordinality constraint on the categories does not allow it.

Figure 6: Same as Figure 5, but with respondents de-emphasized and the positions of the categories for each statement indicated. The label is at the "strongly agree" end of each scale.

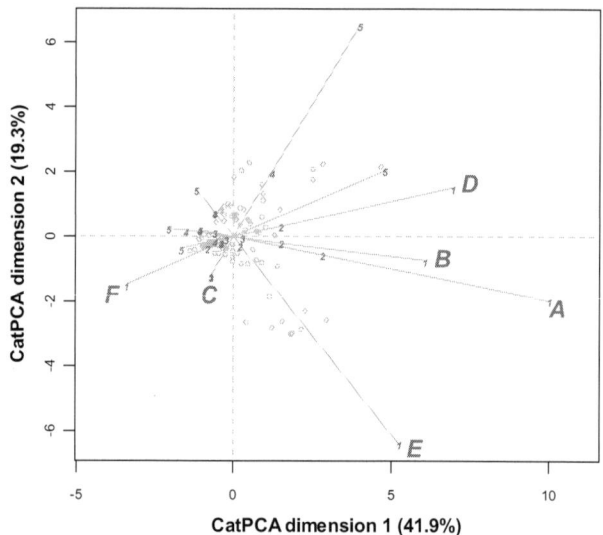

5. The barycentric property of CA

Figures 3 and 4 are biplots. This means that the respondent points can be projected perpendicularly onto the biplot axes defined by the categories (Figure 3) or the categorical variables (Figure 4) and estimates of the data can be deduced. But for Figure 3, this is not interesting as the data are zeros and ones. For Figure 4, this is just like the PCA biplot, with the scale already laid out on the biplot axes. For CatPCA, this projection onto the direction defined by each variable has no real meaning, since the scales are not linear, but the ordering of the projections does have meaning.

There is an alternative relationship between the rows and columns which has meaning across all three biplots, namely the barycentric, or weighted average, relationship. In the MCA of Figure 3, each respondent lies at the (unweighted) average of the six response category points of that correspondent. For example, the respondent with responses [1 1 5 1 2 5] (Figure 3, right hand side) is at the average position of the response category points A1, B1, C5, D1, E2 and F5, which is why it gets its position at the right. The technical reason is that the profile of this respondent is a set of 30 values, all zero except 6 ones at exactly the positions of these categories, with a sum of 6 and thus a profile of 30 values, all zero except the values 1/6 at those response-specific positions. Applying the transition formula in CA, where the profiles values are scalar-product-multiplied with the standard coordinates, this is exactly the ordinary average of the 6 response category points.

Similarly, for the doubled CA, the doubled vector for the same response vector given above is [4 0 | 4 0 | 0 4 | 4 0 | 3 1 | 0 4], with a sum of 24 and thus a profile of [1/6 0 | 1/6 0 | 0 1/6 | 1/6 0 | 3/24 1/24 | 0 1/6] (the order of the doubled categories is A+ A– | B+ B– | etc.). When applied to the standard coordinates of the 12 doubled points, it is a weighted average of them with the profile values as weights. For the responses 1 and 5 (coded 4 0 and 0 4 respectively), these are just putting weight on one of the endpoints, the doubled points, whereas the response 2 (coded as 3 1, for variable E) is placing three times the weight on the E+ end than on the E– at the opposite end. This means that it is placing the weight on the position of response category 2, if we placed a linear scale between E+ (the 1 on the scale) and E– (the 5). So, like MCA, the doubled CA places the respondents at the ordinary average of the 6 response points, where the category points of each variable have been placed at equal intervals on the straight line connecting the two endpoints.

Finally, for CatPCA the result is identical. The scale points are already marked, at unequal intervals between the endpoints (Figure 6), and the respondents find their positions at the averages of the six corresponding response category points.

One aspect of the plotting that should be mentioned here is that Figures 3, 4 and 5 all have the same standard coordinate normalization, for purposes of comparison. This means that the explained variances (i.e., inertias), equal to the average sum of squares of the row points, can be compared across the analyses. They turn out in the expected order, shown here in descending order:

Table 1: Variances (inertias) on the first two dimensions of three alternative analyses of the environment data set

	Dimension 1	Dimension 2
MCA	0.4572	0.3522
CatPCA	0.4194	0.1927
Doubled CA	0.3971	0.1763

MCA, with the least restrictions (and thus with the most free parameters) has the highest explained variances, while doubled CA, with the most restrictions, has the least. CatPCA is between these two because it has the same linear restriction as doubled CA but allows the category points to be anywhere along the line as long as the ordinal property is conserved. Notice the much higher variance of MCA on the second dimension, reflecting the quadratic arch (see Figure 3), which only the MCA can have, and indicating the respondents that have mixed responses on the "optimistic"/"pessimistic" poles.

6. The connection between MCA and PCA

Before considering this connection, the connection between doubled CA and PCA is first summarized, see Greenacre (1984: 175–179). Figure 7 shows the PCA biplot of this data set. The six statements are pointing in almost identical directions compared to the agree sides in Figure 4 with the positive labels. The main difference between the two is that PCA uses the Euclidean distance between the respondents, with variables equally weighted, whereas doubled CA uses the chi-square distance, which is a weighted Euclidean distance, where variables are differentially weighted. Greenacre (1984) showed that these weights depend on the "polarization" of the responses, which is how close the average response is to the extremes. The weight assigned to a variable is least when the average response is at 3, and the weight increases as the average response gets closer to 1 or 5.

Figure 7: PCA of the environmental attitudes data

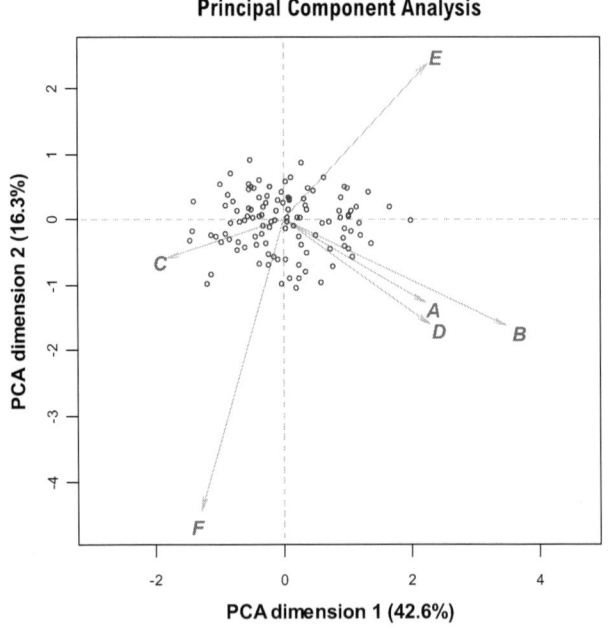

MCA has minimal restrictions on the positions of the category points: only the (weighted) centring of each set of categories at the origin is imposed. Linear restrictions can be imposed to force the categories to lie on a straight line and be at equal intervals. The way to impose these restrictions is given by Golub (1973).

Without entering into all the algebraic details, the way to do this is to set up the linear restrictions on the standard coordinate category positions in p dimensions, say $\mathbf{X}$ (30×p), in the form $\mathbf{CX} = 0$, where $\mathbf{C}$ (r × 30), r being the number of restrictions imposed. As an illustration of one of these restrictions, the category 2 of the first variable must be at a quarter of the distance between categories 1 and 5, that is for the first five rows of $\mathbf{X}$, the second row = first row + ¼ (fifth row − first row). This translates to the first row of $\mathbf{C}$ as [−¾ 1 0 0 ¼ 0 0 ⋯ 0]. There are three restrictions for each variable, totalling r = 18 restrictions. The indicator matrix Z is then projected onto the subspace of these restrictions using the 30 × 30 projection matrix $(\mathbf{I} - \mathbf{V}\mathbf{V}^T)$, where V is the matrix of right singular vectors of C – this is another nice application of the singular value decomposition (SVD). The MCA-style decomposition of $\mathbf{Z}(\mathbf{I} - \mathbf{V}\mathbf{V}^T)$ carries on as before to get the solution of Figure 8, where it can be seen that indeed the categories are on straight lines and at equal intervals.

Figure 8: Linearly constrained MCA of the environmental attitudes data, with differential MCA weights

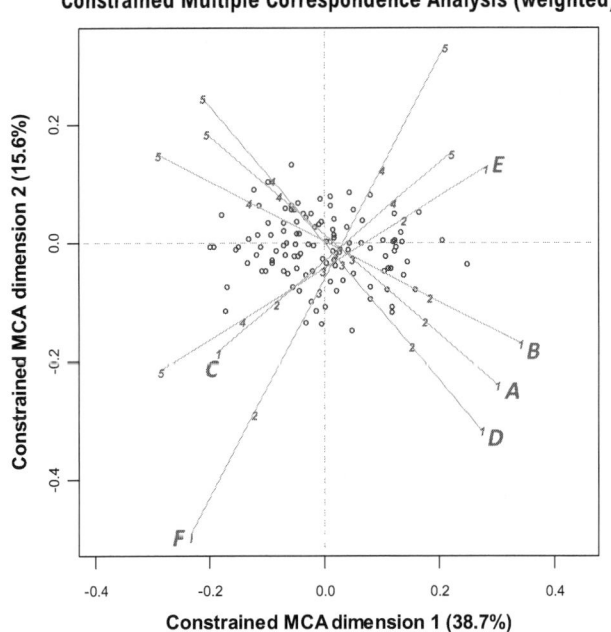

MCA, however, weights each category proportional to its marginal frequency, so in order to arrive at the PCA, equal weights have to be imposed in the MCA. After doing this, the analysis finally arrives at Figure 9, where it can be seen that the middle categories 3 of the scales are all at the centre (cf. Figure 8). The percentages of variance are now exactly the same as the PCA of Figure 7 and the solution is identical apart from an overall scale factor because of the scaling of the MCA compared to the PCA.

As a final comment about the different methods presented here, let us note the number of dimensions with positive variance in each of the methods. Starting with MCA, it is known that there are $J - Q = 30 - 6 = 24$ dimensions: 6 out of the 30 are set to zero because of the 6 centring restrictions on the variables. At the other extreme, a doubled CA and a PCA have only 6 dimensions: for the 6-column PCA this is clear, while for the 12-column doubled CA the pairs of columns are linearly related, thus reducing the dimensionality from 12 to 6. For the CatPCA, the imposition of linearity on the 6 variables seems to imply reducing the MCA dimensionality by 6, since there are still three free scale positions per variable, hence dimensionality 18. In the PRINCALS implementation of CatPCA (Mair and de Leeuw 2022) in R (R Core Development Team 2022), only 6 dimensions are given, since each categorical variable is reduced to a single quantified one. In the constrained MCA with weights of Figure 8 the dimensionality is 12, since the endpoints are free and the dimensionality is reduced by 18 by the restrictions. Finally, the unweighted constrained MCA of Figure 9 has only 6 dimensions, just like the PCA, since the 6 variables are additionally constrained to have their middle categories at the origin (this could have been equivalently achieved by adding six more restrictions where the endpoints are the reverse sign of each other.

Figure 9: Linearly constrained MCA of the environmental attitudes data, with equal category weights. Apart from an overall scaling factor, this is now identical to the PCA of Figure 7

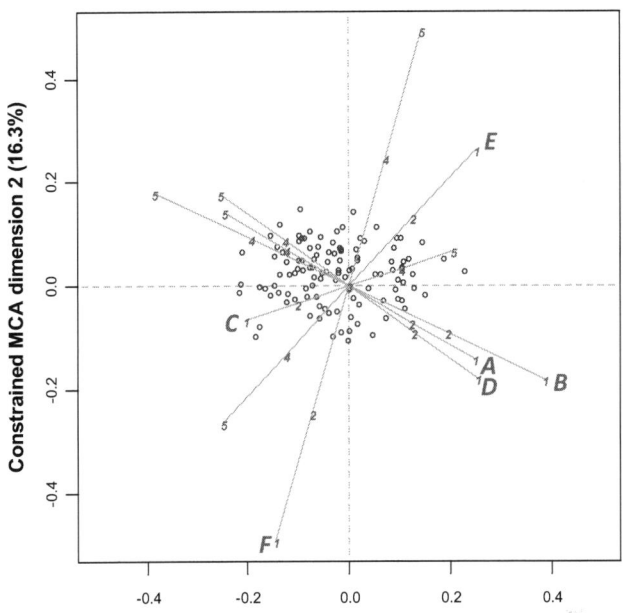

7. Conclusion

CA and PCA share a common theory, but have inherent differences and fields of application. An understanding of the connections that have been shown in this paper gives additional insight into the workings of these two methods and how they should be interpreted.

References

Aşan, Zerrin/Greenacre, Michael (2011): Biplots of fuzzy coded data. Fuzzy Sets and Systems:183, 57–71.

De Leeuw, Jan (2014): History of nonlinear principal component analysis. In: Blasius, Jörg/Greenacre, Michael (eds.): The Visualization and Verbalization of Data. Boca Raton, Florida: Chapman & Hall / CRC Press.

Golub, Gene (1973): Some modified matrix eigenvalue problems. SIAM Review: 15, 318–334.

Greenacre, Michael (1984): Theory and Applications of Correspondence Analysis. Paris: Dunod.

Greenacre, Michael (1988): Correspondence analysis of multivariate categorical data by weighted least squares, Biometrika: 75, 457–467.

Greenacre, Michael (2016): Correspondence Analysis in Practice, Third Edition. Boca Raton, Florida: Chapman & Hall / CRC Press.

Greenacre, Michael (2017): 'Size' and 'shape' in the measurement of multivariate proximity. Methods in Ecology and Evolution: 8: 1415–1424. https://doi.org/10.1111/2041-210X.12776

Greenacre, Michael (2018): Compositional Data Analysis in Practice. Boca Raton, Florida: Chapman & Hall / CRC Press.

Greenacre, Michael/Groenen, Patrick J.F./Hastie, Trevor/Iodice d'Enza, Alfonso/Markos, Angelos/Tuzilhina, Elena (2022): Principal component analysis. Nature Reviews Methods Primer: 2, 101. https://doi.org/10.1038/s43586-022-00192-w

ISSP Research Group (2020): International Social Survey Programme: Environment IV – ISSP 2020. ZA7650 Data file. Cologne: GESIS.

Mair, Patrick/De Leeuw, Jan (2022): Gifi: Multivariate Analysis with Optimal Scaling. R package version 0.4-0/r383. https://R-Forge.R-project.org/projects/psychor/

R Core Team (2022): R: A language and environment for statistical computing. R Foundation for Statistical Computing, Vienna, Austria. https://www.R-project.org/

Dr. Strangelove or: How I Learned to Stop Worrying and Love the Correspondence Analysis

Jörg Breitung

Abstract

This paper analyses the properties of alternative distance metrics for analysing contingency tables. It is argued that the usual χ^2 -metric possesses some undesirable features and it is therefore interesting to consider alternative metrics. Furthermore it is shown that the "standardized residuals" used as a starting point for the correspondence analysis are not properly standardized as their variances depend on cell probabilities. As an alternative, the deviation from independence is measured by the underlying correlation coefficient. It is argued that this measure is similar to the (signed square-root) of the log-likelihood difference, suggesting that this metric shares some more appealing statistical properties.

1. Introduction

What do you do, if you don't know anything about something that you do not really care about? Right, you ask Google. So I typed: "What is correspondence analysis"? As usual, one of the first entries of the search results is Wikipedia. I read: "*Correspondence Analysis is conceptually similar to principal component analysis, but applies to categorical rather than continuous data.*" Oh that's great because I have some experience with principal component analysis (see e.g. Breitung and Eickmeier 2011, Breitung and Tenhofen 2011, Breitung 2013). I therefore started to study the relevant textbooks like Blasius (2001) and Greenacre (2007) and try to understand the relationship between correspondence analysis (*henceforth: CA*) and principal component analysis (PCA). Obviously the literature on CA is huge and for this small note I was not able to carefully study the relevant literature. So I stick to the position of an alien that is visiting the planet "*correspondence analysis*". This alien has grown up on the planet called "Econometrics", where data matrices are typically pretty large and observed on metric scales. Utmost importance is attached to define and identify what you are estimating, and no one will take you seriously unless you can demonstrate that your results are statistically significant and that the assumptions of your model are valid.

This said I carefully approached the CA. If you do so, you first have to survive an exhausting trip through a (non)Euclidean geometric space, which is somewhat unusual in statistical analysis. To be honest, this geometric journey has left me a bit perplexed. To understand better the geometric approach, I first studied the geometry of CA and found that the χ^2-distance has some undesirable features and it may therefore be appealing to consider alternative metrics as well. My favorite is to transform the entries of the contingency table into correlation coefficients. Interestingly, correlations are constructed similarly as the χ^2-distance but imply a slightly different denominator. Another measure is the likelihood ratio (LR) that seems most natural from a statistical point of view. Moreover, LR statistics allow us to find out whether some specific cell frequency (or row/column) is significantly different from the expected cell frequency under the assumption of independent outcomes.

To illustrate the issues involved, I borrowed the empirical example of Blasius and Greenacre (2006) and compare different distance measures and visualization techniques. In particular, I present the correlation matrix in form of a "balloon plot" that highlights the distances from the expected cell frequencies. I argue that the information presented by the balloon plot is similar to the biplot in CA but it does not involve a loss of information from dimensionality reduction. Furthermore I visualize the correlation matrix by a respective biplot that reveals interesting differences to the standard correspondence plot in CA.

The rest of the paper is organized as follows. In Section 2 I compare PCA and CA thereby highlighting some conceptual differences between these approaches. The standard χ^2-distance measure is considered in Section 3. Following Rao (1995) I argue that this distance measure possesses some undesirable properties. Another drawback is that under the assumptions of independent outcomes the variances of the χ^2-distances depend on the cell probabilities and, therefore, the entries of the contingency matrix are not properly standardized. To overcome these disadvantages, Section 4 proposes a distance measure constructed from the underlying correlation coefficient that also provides a proper standardization. In Section 5 I consider the standard statistical distance measure namely the difference of the log-likelihood function (resp. the logarithm of the likelihood ratio, LR). It turns out that the LR distance is related but not identical to the χ^2-distance. By construction, the LR distances shares some optimality properties for statistical inference. The theoretical results are illustrated in Section 6 by using the empirical example of Blasius and Greenacre (2006). Section 7 offers some concluding remarks.

2. The relationship between CA and PCA

PCA is typically applied to large correlation matrices. Let X denote an $N \times J$ matrix where N indicates the number of observations on each of the J variables. The data is standardized such that the diagonal elements of the matrix $N^{-1}X'X$ are equal to one. The PCA is a dimension reduction technique that maps the J variables in a subspace spanned by $J \gg r$ linear combinations of the original variables (the so-called principal components).

Whereas PCA considers (standardized) observations from a large number of variables, the CA starts from a matrix constructed from a particular distance metric (see Section 3). Furthermore, in most textbook examples and empirical applications the size of the contingency table is rather small, with K and L typically less than 10. In such situations it is not obvious that it is necessary to reduce the dimensionality. As far as I can see the main reason for reducing the dimension of the correspondence matrix is the possibility to represent the data in form of a two-dimensional diagram (biplot). Obviously, such a dimensionality reduction implies some loss of information but helps to recover some unknown structure behind the contingency matrix.

Typically, the data for a $K \times L$ contingency table come in form of n bivariate vectors $(a_1,b_1),(a_2,b_2),\ldots,(a_n,b_n)$ of independent multinomially distributed random variables, where $a_i \in \{1,2,\ldots,K\}$, $b_i \in \{1,2,\ldots,L\}$, and $i = 1,\ldots,n$.

Let us define $K + L$ corresponding dummy variables as

$$d_i^a(k) = \begin{cases} 1 & \text{if } a_i = k \\ 0 & \text{otherwise} \end{cases}$$

$$d_i^b(\ell) = \begin{cases} 1 & \text{if } b_i = \ell \\ 0 & \text{otherwise} \end{cases}$$

Then, the relative frequency results as

$$p_{k\ell} = \frac{n_{k\ell}}{n} = \frac{1}{n}\sum_{i=1}^{n} d_i^a(k)d_i^b(\ell),$$

where $n_{k\ell}$ indicates the number of observations in the (k,ℓ)-cell of the contingency table. The difference between actual and expected (by assuming independence) relative cell frequencies result as

$$q_{k\ell} := p_{k\ell} - p_{k\cdot}p_{\cdot\ell} = \frac{1}{n}\sum_{i=1}^{n} d_i^a(k)d_i^b(\ell) - \left(\frac{1}{n}\sum_{i=1}^{n} d_i^a(k)\right)\left(\frac{1}{n}\sum_{i=1}^{n} d_i^b(\ell)\right),$$

where $p_{k\cdot} = \sum_{\ell=1}^{L} p_{k\ell}$ and $p_{\cdot\ell} = \sum_{k=1}^{K} p_{k\ell}$ are called the row and column masses, respectively. This shows that the "residual" $q_{k\ell}$ can be interpreted as the sample covariance between the dummy $d_i^a(k)$ variables and $d_i^b(\ell)$. In Section 4 this interpretation is employed to construct a correlation matrix between the outcomes of the contingency table.

For the CA we divide the residuals $q_{k\ell}$ by the factor $\sqrt{p_{k\cdot}p_{\cdot\ell}}$ yielding what is often called the "standardised residuals" (e.g. Blasius 2001: 89, and Greenacre 2007: 202). In Section 3 I argue that dividing the residuals by $\sqrt{p_{k\cdot}p_{\cdot\ell}}$ does not result in a proper standardization, as the variances of the χ^2-distances depend on the cell probabilities.

Let S denote the $K \times L$ matrix with typical element

$$s_{k\ell} = \frac{q_{k\ell}}{\sqrt{p_{k\cdot}p_{\cdot\ell}}} = \frac{p_{k\ell} - p_{k\cdot}p_{\cdot\ell}}{\sqrt{p_{k\cdot}p_{\cdot\ell}}}. \tag{1}$$

The reduction of dimensionality is obtained by applying the singular value decomposition (SVD) with $S = UDV'$. Let $U_2 (V_2)$ denote the matrix of the first two left (right) singular vectors and D_2 is the upper-left (2×2) submatrix of D. In PCA the biplot coordinates are given by $U_2 D_2^\alpha$ and $V_2 D_2^{1-\alpha}$ where an asymmetric version is most popular by choosing either $\alpha=0$ or $\alpha=1$. The set of points depicting the variables is typically drawn as arrows from the origin to reinforce the idea that they represent biplot axes onto which the observations can be projected when approximating the original data.

The SVD results in a least-squares minimal approximation of the element in S given by

$$s_{k\ell} = u'_{2k} D_2 v_{2\ell} + \tilde{e}_{k\ell}$$

where u_{2k} $(v_{2\ell})$ represents the k-th (ℓ-th) row of $U_2 (V_2)$ and $\tilde{e}_{k\ell}$ is the approximation error. So far so comprehensible. But now correspondence analysis introduces another transformation yielding

$$\frac{1}{\sqrt{p_{k\cdot}p_{\cdot\ell}}} s_{k\ell} = \frac{p_{k\ell}}{p_{k\cdot}p_{\cdot\ell}} - 1 = \left(\frac{u'_{2k}}{\sqrt{p_{k\cdot}}}\right) D_2 \left(\frac{v_{2\ell}}{\sqrt{p_{\cdot\ell}}}\right) + \frac{1}{\sqrt{p_{k\cdot}p_{\cdot\ell}}} \tilde{e}_{k\ell}$$
$$:= \phi'_k D_2 \gamma_\ell + e_{k\ell}$$

where ϕ_k and γ_ℓ are called "standard coordinates" (cf. Greenacre 2007: 202). It is important to note, however, that the error $e_{k\ell}$ is no longer a least-squares minimal approximation error. Instead, the least-squares minimal approximation is obtained from the SVD applied to the matrix $D_r^{-1/2} S D_c^{-1/2}$, here D_r and D_c are diagonal matrices with the row and column masses $(p_{k\cdot}$ resp. $p_{\cdot\ell})$ on the main diagonals. The latter SVD yields a different representation with smaller approximation error. As an example consider the Asbestos data set (see Selikoff 1981).[1] The two alternative approaches for computing the standard coordinates yield the following matrices of standard row coordinates:

$$D_r^{-1/2}U_2 = \begin{pmatrix} -0.847 & 0.472 \\ 0.416 & -1.334 \\ 1.800 & 0.879 \\ 2.161 & 2.167 \end{pmatrix}, \quad \tilde{\Phi} = \begin{pmatrix} -0.224 & 0.476 \\ 0.045 & -0.689 \\ 0.582 & -0.338 \\ 0.780 & 0.429 \end{pmatrix}$$

where $\tilde{\Phi}$ is obtained as the two left eigenvectors of the SVD for $D_r^{-1/2}SD_c^{-1/2}$.[2]

This alternative ("direct") way of computing the standard coordinates is also mentioned by Greenacre (2007, eq. (A.13)). What is obvious from the above results is that the two approaches may render quite different coordinates.

An important issue in correspondence analysis is the proper scaling of the biplot axes. The R package "ca" (cf. Nenadić and Greenacre 2007) offers 8 different scaling options for the coordinates. Obviously, the issue of the "best scaling" of the coordinates is not settled completely. The standard scaling employs *principal coordinates* given by $F=D_r^{-1/2}U_2D_2$ and $G=D_c^{-1/2}V_2D_2$. This particular scaling is chosen such that the weighted sum-of-squares of the principal coordinates (i.e. their inertia in the direction of this dimension) is equal to the square of the singular value (the principal inertia).

Summing up, PCA and CA share the idea of representing some matrix by a lower dimensional approximation. But the details of the analysis are quite different. Statistical inference using PCA typically assumes an i.i.d. sample of J correlated variables, where $r \ll J$ linear combinations of the variables (principal components) are constructed that best represent the linear dependence among the variables.[3] On the other hand, the CA analyses the variability of the rows/columns of the contingency table. Notwithstanding these conceptual differences the CA benefits from adapting useful tools like the biplot in order to visualize the data.

---

1 The R code for this and the other computations are provided on the homepage of the author.
2 The coordinates are available from the output ($rowcoord) of the R package "ca".
3 The i.i.d. assumption may be dropped by allowing for some "weak correlation" (e.g. Bai 2003) but the data matrix of the CA may be strongly correlated in both dimensions

3. The χ^2-distance

Textbooks on CA (e.g. Blasius 2001 and Greenacre 2007) typically start with some geometric reasoning in Euclidean space in order to explain how to measure the distance between the row resp. column profiles. Then Pearson's χ^2-statistic for independence is introduced and it is argued that this test statistic gives rise to the χ^2-distance measure for two row profile vectors r_k and $r_{k'}$ defined as (cf. Greenacre 2007: 31)

$$\| r_k - r_{k^*} \|_\chi = \sqrt{\sum_{\ell=1}^{L} \frac{\left[(p_{k\ell}/p_{k\cdot}) - (p_{k^*\ell}/p_{k^*\cdot})\right]^2}{p_{\cdot\ell}}}$$

$r_k = (p_{k1}/p_{k\cdot}, \ldots, p_{kL}/p_{k\cdot})'$ and r_{k^*} denote two $L \times 1$ vectors of row profiles. The χ^2-statistic results as a weighted average of the row (resp. column) distances

$$X^2 = n \sum_{k=1}^{K} p_{k\cdot} \| r_k - c \|_\chi^2$$

where $c = (p_{\cdot 1}, \ldots, p_{\cdot L})'$ is the average row profile under the assumption of independence (i.e. the vector of column masses). A similar representation can be derived for the column distances.

As far as I can see, the main reason for introducing such a distance metric is the desire to develop a geometric interpretation for Pearson's χ^2 statistic. This is achieved by defining the χ^2 distance as a *weighted* Euclidean distance. For me as an alien it is difficult to see why the squared distances should be weighted by the (inverted) column masses. The only reason seems to be that applying this particular weighting scheme gives rise to the χ^2 statistic. But why should the distance between the entries of two rows profiles depend on the respective column masses? This implies that when we add or drop rows, then the corresponding distances may get smaller or larger. Another problem is that if the column masses get small, then undue emphasis is given to the corresponding row distances. This led the famous (and now 102 years old) C.R. Rao (1995) to advocate the Hellinger distance defined as

$$d_H(r_k, r_{k^*})^2 = \sum_{\ell=1}^{L} (\sqrt{p_{k\ell}/p_{k\cdot}} - \sqrt{p_{k^*\ell}/p_{k^*\cdot}})^2$$

Beside the fact that this distance measure only depends on the row profiles themselves and does not imply any weighting related to information from outside, this distance measure satisfies the principle of *distributional equivalence* and the distances do not become arbitrarily large if the column masses tend to zero. Cuadras and Cuadras (2006) proposed a generalized distance measure that entails the CA distances and the Hellinger distance as a special case. Other alternatives are the L_1-type distance of Benzécri (1982) and the log-ratios considered in Cuadras and Cuadras (2006). This list of proposed distance measures is not complete. Consider, for example, a distance measure that is popular in machine learning when it comes to analyzing the similarity of discrete distributions[4] (e.g. Yang et al. 2015), which is defined as

$$\| r_k - r_{k*} \|_S = \sqrt{\sum_{\ell=1}^{L} \frac{(p_{k\ell} - p_{k*\ell})^2}{2(p_{k\ell} + p_{k*\ell})}} \qquad (2)$$

Note that under independence we have $p_{k\ell} \approx p_{k.} p_{.\ell}$ and, therefore, for independent outcomes this we have

$$4n \sum_{k=1}^{K} \| r_k - c \|_S^2 \approx X^2.$$

Let us now consider the Pearson's statistic given by

$$X^2 = n \sum_{k=1}^{K} \sum_{\ell=1}^{L} s_{k\ell}^2$$

where $S_{k\ell}$ as defined in (1) is often called the "standardized residuals". A proper standardization would imply that $S_{k\ell}$ has the same (unit) variance for all k and ℓ but it turns out that the distributional properties of $S_{k\ell}$ depend on the cell probabilities. The reason is that the denominator $\sqrt{p_{k.} p_{.\ell}}$ is different from the standard deviation of the numerator, even if the outcomes are independent. To illustrate this fact I performed a small Monte Carlo experiment. The data generating process resembles the Asbestos data set used in the previous section, where the data is generated as independent multinomial distributed random variables with probabilities $p_{k.} p_{.\ell}$, that is, the data are generated under the assumption of independent outcomes. The results of the Monte Carlo simulation based on 1000 replications are presented in Table 1.

---

4   See also the function *chisqDistance(a,b)* in the R package *colordistance* (e.g. Yang et al. 2015),

Table 1: Variances of the residuals with different standardization

row	$p_{k\ell}$	var($S_{k\ell}$)	var($\varrho_{k\ell}$)	$p_{k\ell}$	var($S_{k\ell}$)	var($\varrho_{k\ell}$)
		first column			second column	
1	0.159	0.346	1.036	0.101	0.464	1.000
2	0.174	0.328	1.025	0.111	0.451	1.015
3	0.035	0.471	1.045	0.022	0.610	0.975
4	0.089	0.395	0.987	0.056	0.557	1.003
5	0.055	0.427	0.988	0.035	0.592	0.988
row		third column			fourth column	
1	0.034	0.597	0.975	0.013	0.629	0.954
2	0.038	0.616	1.051	0.015	0.631	1.001
3	0.007	0.853	1.033	0.003	0.904	1.017
4	0.019	0.764	1.042	0.007	0.772	0.978
5	0.012	0.826	1.044	0.004	0.805	0.945

Entries present the variances of the standardized residuals as defined in (1) and (3). Entries are based on 1000 replications of the Asbestos dataset generated under the assumption of independent outcomes.

The results indicate that $S_{k\ell}$ cannot be considered to be properly standardized as the variances tend to become larger for smaller cell probabilities $p_{k\ell}$. The variances range from 0.328 for a cell probability of 0.174 up to 0.904 for a probability of 0.003. Accordingly, it is much more likely to observe large residuals when the corresponding probability is small. On the other hand, the standardisation proposed in the next section appear to work well (indicated by $\varrho_{k\ell}$ in Table 1).

4. Correlation as a distance measure

There is a close relationship between distance and correlation. For example, consider the Euclidean distance

$$\|x-y\| = \sqrt{\sum_{i=1}^{n}(y_i - x_i)^2}$$

between two vectors of standardized random variables $x = (x_1, \ldots, x_n)'$ and $y = (y_1, \ldots, y_n)'$. Since

$$\|x-y\|^2 = 2n - 2\sum_{i=1}^{n} x_i y_i$$

the squared Euclidean distance can be expressed as $\|x-y\|^2 = 2n(1 - \varrho_{xy})$, where ϱ_{xy} denotes the correlation coefficient between x_i and y_i. It therefore makes sense to consider the relationship between distances and correlations between the outcomes of the contingency table.

As noted in Section 2, the differences $p_{k\ell} - p_i p_{\cdot j}$ can be written as the covariance between the two dummy variables $d_i^a(k)$ and $d_i^b(\ell)$. This suggests to use the correlation between the dummy variables as measure of the deviation from independence. Since the dummy variables are binomially distributed with variances $var[d_i^a(k)] = p_{k\cdot}(1 - p_{k\cdot})$ and $var[d_i^b(l)] = p_{\cdot l}(1 - p_{\cdot l})$ and therefore, an estimator for the correlation between the two dummy variables results as

$$\varrho_{k\ell} = \frac{p_{k\ell} - p_i p_{\cdot j}}{\sqrt{(p_{k\cdot} - p_{k\cdot}^2)(p_{\cdot j} - p_{\cdot j}^2)}} \qquad (3)$$

It is interesting to note the close correspondence between this correlation measure and the standardized residuals $S_{k\ell}$ used for the correspondence analysis. The only difference is the squared probabilities in the denominator. Since the cell probabilities are typically small, the differences between $S_{k\ell}$ and $\varrho_{k\ell}$ are usually moderate. An important property of the correlation coefficient is that for independent outcomes

$$\sqrt{n}\varrho_{k\ell} \xrightarrow{d} N(0,1)$$

where $\xrightarrow{d}$ signifies convergence in distribution. Accordingly, under the hypothesis of independent outcomes all correlations have the same asymptotic distribution and can therefore be considered to be properly standardized. In Section 6 I therefore use the correlations $\varrho_{k\ell}$ instead of $s_{k\ell}$ as an alternative starting point of the correspondence analysis.

5. The log-likelihood distance

It is well known that for statistical tests the difference between the log-likelihood functions under the null and alternative hypotheses results in most powerful test statistics (Neyman-Pearson lemma). It is therefore natural to consider the difference in the log-likelihood functions (that is the logarithm of the likelihood-ratio) as a distance measure between the actual and expected cell frequencies.

Let us therefore consider the log-likelihood difference for a test of the hypothesis of independent outcomes which given by

$$LR = 2n \sum_{k=1}^{K} \sum_{\ell=1}^{L} p_{k\ell} \log\left(\frac{p_{k\ell}}{p_{k\cdot} p_{\cdot \ell}}\right). \qquad (4)$$

How is this likelihood-ratio statistic related to Pearson's χ^2-statistic? Let $p_{k\ell}^0 = p_{k\cdot} p_{\cdot \ell}$.

A second order Taylor expansion around $p_{k\ell}^0$ yields:

$$p_{k\ell}\left[\log(p_{k\ell}) - \log(p_{k\ell}^0)\right] \approx (p_{k\ell} - p_{k\ell}^0) + \frac{1}{2p_{k\ell}^0}(p_{k\ell} - p_{k\ell}^0)^2.$$

Since $\sum_{k=1}^{K}\sum_{\ell=1}^{L}(p_{k\ell} - p_{k\ell}^0) = 0$ it follows that the likelihood ratio statistic LR can be approximated by Pearson's χ^2 statistic whenever $p_{k\ell} - p_{k\ell}^0$ is small (that is, if the outcomes are nearly independent).

Let us now consider the LR test for the hypothesis that a particular row profile, say for $k = 1$, deviates from the other row profiles. It is important to notice that it is not possible to just pick the relevant summands for $k = 1$ from the LR statistic (4) as this would not result in a test statistic with the usual χ^2-distribution with $L - 1$ degrees of freedom. Instead we consolidate the contingency table such that

$k \downarrow$	$\ell : 1$	2	...	L
1	p_{11}	p_{12}	...	p_{1K}
2	$\bar{p}_{21}$	$\bar{p}_{22}$	...	$\bar{p}_{2K}$

Where $\bar{p}_{2\ell} = \sum_{k=2}^{K} p_{k\ell}$. Accordingly, the remaining rows $k = 2, 3, \ldots, K$ are aggregated such that the contingency table is reduced to a $2 \times L$ table. The LR statistic for this row results as

$$LR(k=1) = 2n\sum_{\ell=1}^{L} p_{1j}\log\left(\frac{p_{1j}}{(p_{1\ell} + \bar{p}_{2\ell})p_{1\cdot}}\right) + p_{1j}\log\left(\frac{\bar{p}_{2\ell}}{(p_{1\ell} + \bar{p}_{2\ell})\bar{p}_{2\cdot}}\right) \quad (5)$$

where $\bar{p}_{2\cdot} = \bar{p}_{21} + \bar{p}_{22} + \cdots + \bar{p}_{2L}$. This test statistic is asymptotically χ^2-distributed with $L - 1$ degrees of freedom. In what follows I refer to this LR statistic as the *LR row distance*. In this manner we can also define a log-likelihood distance for each cell. To this end we need to consolidate also the remaining columns such that for the upper right cell, for example, we obtain

$k \downarrow$	$\ell : 1$	2
1	p_{11}	$\bar{p}_{12}$
2	$\bar{p}_{21}$	$\bar{p}_{22}$

where $\bar{p}_{12} = \sum_{\ell=2}^{L} p_{1\ell}$, $\bar{p}_{21} = \sum_{k=2}^{K} p_{k1}$ and $\bar{p}_{22} = \sum_{k=2}^{K}\sum_{\ell=2}^{L} p_{k\ell}$. The log-likelihood difference for testing independence in this 2×2 matrix is denoted by LR($k = 1$, $\ell = 1$). Let us now consider the properties of this distance measure as an alternative to the χ^2-distance. First, it is obvious that under independence the log-likelihood difference is asymptotically χ^2-distributed whereas the distribution of the χ^2-distance depends on the cell probabilities. Furthermore it turns out that LR$(k, \ell) \approx n\varrho_{k\ell}^2$ and, therefore, we may use the signed square-root of the LR(k, ℓ) statistic as an alternative for constructing properly standardized residuals.

6. An empirical illustration

To illustrate the issues discussed in the previous sections, I borrow an example from Blasius and Greenacre (2006). The data are from the International Social Survey Program (ISSP). Table 2 reports responses from five selected countries to the question: *"When my country does well in international sports, it makes me proud to be [Country Nationality]"*. The contingency table is provided in Table 2.

Table 2: Contingency table for "international sports × countries"

	U.K.	U.S.	Russia	Spain	France
agree strongly	230	400	1010	201	365
agree	329	471	530	639	478
neither nor	177	237	141	208	305
disagree	34	28	21	72	50
disagree strongly	6	12	11	14	97

	Column profiles					
	U.K.	U.S.	Russia	Spain	France	row masses
agree strongly	0.296	0.348	0.590	0.177	0.282	0.364
agree	0.424	0.410	0.309	0.563	0.369	0.403
neither nor	0.228	0.206	0.082	0.183	0.236	0.176
disagree	0.044	0.024	0.012	0.063	0.039	0.034
disagree strongly	0.009	0.010	0.006	0.012	0.075	0.023
column masses	0.128	0.189	0.282	0.187	0.213	
LR difference	37.41	21.98	568.00	265.68	260.89	

Source: Blasius and Greenacre (2006: 7). "LR difference" indicates the LR statistic for the hypothesis that the respective row profile is identical to the expected profile under independence.

Let us first address the question: how special are the responses from the different countries? To assess the deviation of each column (country) to all other countries, I computed the log-likelihood differences as in (4) after transposing the matrix to obtain the LR column distances. The results are presented in the

39

last row of Table 2. These results suggest that the responses of Russia are most different from the responses of the other countries, whereas the responses of Spain and France are less different but still quite far away from the "average column" (row masses). The columns of the U.K. and U.S. are much more similar to the average (resp. independent) pattern. We can also compare the outcomes of two countries with each other. For example the log-likelihood difference between the U.K. and U.S. is only 10.834, suggesting that the response pattern of these countries are pretty similar. On the other hand, the log-likelihood difference between the columns of France and Spain is 145.06. Although the distance of these two columns to the mean column is similar, this does not mean that the response pattern of France and Spain is similar. This becomes also clear from the correspondence analysis (see below).

Table 3 compares the standardized residuals $s_{k\ell}$ to the alternative measures considered in Sections 3 – 5. The first line for each row presents the respective cell entries of the matrix S as defined in (1), whereas the second line reports the symmetric distances measure $\| \cdot \|_s$ as defined in (2). In most cases the differences between these two distance measures are pretty small. The correlation measure reveals more important differences to the χ^2-distance measures in particular if the distance gets large. For example for the (1, 3)-cell the χ^2-distance is 0.199, whereas the correlation is 0.294. This suggests that larger distances are more accentuated by using correlations instead of χ^2-distances. By multiplying correlations with $\sqrt{n}$ we obtain a test statistic for the null hypothesis that the two features affecting the cell outcome are independent. This hypothesis is rejected at the 0.05 significance level for 15 out of 25 cells.

Table 3: Alternative distance measures

(row,column):	(1,1)	(1,2)	(1,3)	(1,4)	(1,5)
s_{kl} (residual)	-0.039	-0.010	0.199	-0.133	-0.062
Symmetric	-0.041	-0.011	0.174	-0.155	-0.066
Correlation	-0.053	-0.015	0.294	-0.185	-0.088
$\sqrt{n}$.corr.	-4.171	-1.191	22.94	-14.47	-6.900
LR distance	-4.226	-1.194	22.68	-15.13	-7.000
(row,column):	(2,1)	(2,2)	(2,3)	(2,4)	(2,5)
s_{kl} (residual)	0.011	0.004	-0.078	0.108	-0.024
Symmetric	0.011	0.004	-0.083	0.100	-0.025
Correlation	0.016	0.006	-0.120	0.156	-0.036
$\sqrt{n}$.corr.	1.250	0.527	-9.361	12.18	-2.835
LR distance	1.248	0.527	-9.459	12.07	-2.846
(row,column):	(3,1)	(3,2)	(3,3)	(3,4)	(3,5)
s_{kl} (residual)	0.044	0.031	-0.118	0.007	0.065
Symmetric	0.041	0.030	-0.138	0.007	0.060
Correlation	0.052	0.038	-0.154	0.009	0.081
$\sqrt{n}$.corr.	4.074	3.001	-12.02	0.721	6.334
LR distance	3.957	2.953	-12.77	0.718	6.161
(row,column):	(4,1)	(4,2)	(4,3)	(4,4)	(4,5)
s_{kl} (residual)	0.019	-0.022	-0.062	0.069	0.012
Symmetric	0.018	-0.023	-0.075	0.058	0.012
Correlation	0.021	-0.025	-0.074	0.078	0.013
$\sqrt{n}$.corr.	1.654	-1.958	-5.822	6.137	1.081
LR distance	1.594	-2.036	-6.394	5.661	1.064
(row,column):	(5,1)	(5,2)	(5,3)	(5,4)	(5,5)
s_{kl} (residual)	-0.036	-0.036	-0.058	-0.030	0.157
Symmetric	-0.044	-0.042	-0.073	-0.035	0.108
Correlation	-0.039	-0.040	-0.069	-0.034	0.179
$\sqrt{n}$.corr.	-3.048	-3.164	-5.420	-2.669	14.00
LR distance	-3.469	-3.462	-6.085	-2.871	12.33

In order to visualize the correlation pattern, Figure 1 presents a balloon plot for the correlations. The larger the positive correlation the larger is the green (resp. black) balloon, whereas negative correlations are indicated by red (grey) balloons. As the more popular alternative Figure 2 presents the correspondence plot as provided by the R package "ca" (Nenadić and Greenacre 2007).[5] Let me summarize the main findings of Blasius and Greenacre (2006: 10f) from analysing this contingency table in their own words followed by the corresponding pattern in the balloon plot:

> "The first dimension can be interpreted as 'level of pride towards achievement in international sport. ... As for the countries we see Russia on the left opposing the other countries of the right; thus of these five nations the Russians feel most proud when Russia is doing well in international sports. At the opposite right-hand side of this axis we see that the French and the Spanish are the least proud of the five nations in this respect...".

This conclusion is confirmed by the balloon plot (Figure 1). The nationality Russian is most correlated with strongly agree, suggesting that Russians feel most proud for the achievements in international sports. By contrast, the category "strongly agree" is highly negatively correlated with "France" and "Spain" which corresponds with the findings of the correspondence plot, where these countries are far away from the category "strongly agree".

> "The second dimension mainly reflects the outlying position "disagree strongly" as well as France compared with the other categories and countries."

Indeed this "outlier' is represented by the high positive correlation (green balloon) between "disagree strongly" and France.

> "...the U.S. and U.K. have very similar response pattern, which are not much different from the overall, or average, pattern. Geometrically this is depicted by these two countries lying close to each other, towards the origin of the map."

In the balloon plot this fact can be identified by the rather small correlations for all categories of these two countries.

---

5   The plot in figure 2 is slightly different from the CA plot in Blasius and Greenacre (2006), as the first axis seems to be multiplied by -1. Note that the sign of the axes is not identified.

Figure 1: Balloon plot of correlations

	U.K.	U.S.	Russia	Spain	France
agree strongly	●	·	⬤	⬤	●
agree	·	·	⬤	⬤	●
neither nor	●	●	⬤	·	●
disagree	·	●	●	●	·
disagree strongly	●	●	●	●	⬤

Note: Green (resp. black) balloon: positive correlation, red (grey) balloon: negative correlation. The size of the balloon corresponds to the absolute correlation between the dummy variables indicating the (i, j) cell of the contingency table.

Figure 2: Symmetric correspondence plot

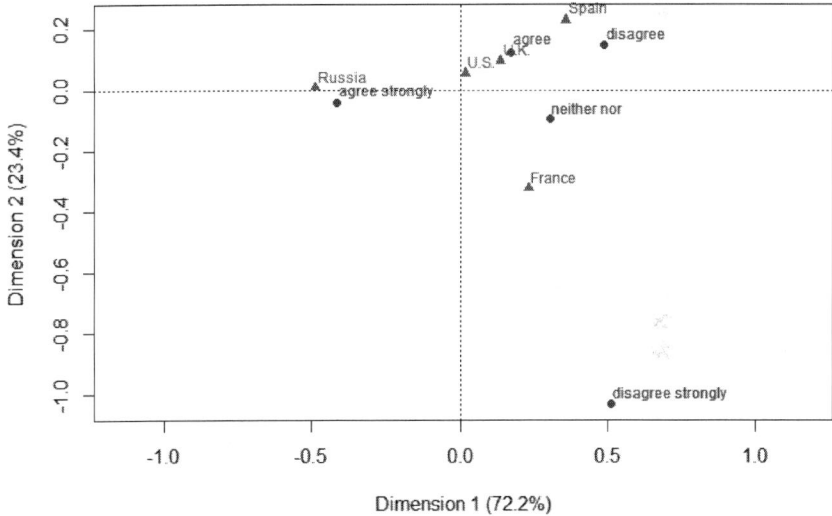

Note: Correspondence plot as obtained from the R package "ca".

43

Figure 3: Symmetric biplot of correlations

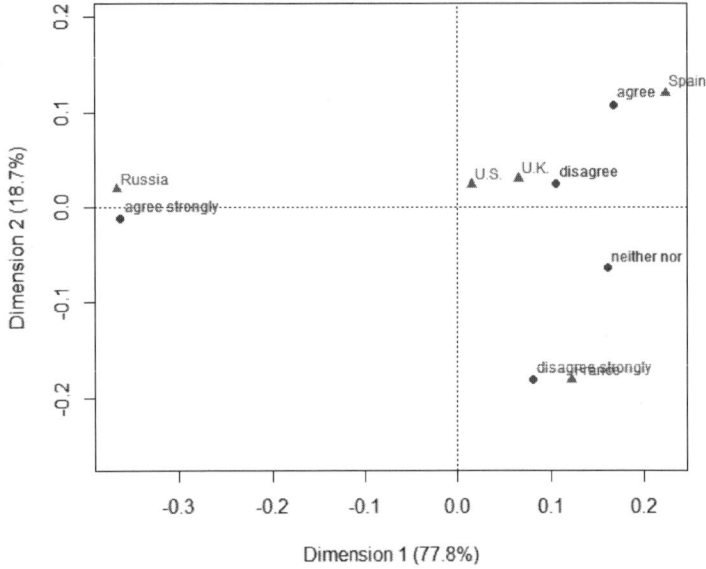

Note: Modified correspondence plot based on correlations.

It appears that many features of the correspondence plot can also be obtained from studying the correlations presented by the balloon plot. Categories that are close together in the correspondence plot are highly correlated, whereas observations far away from each other are negatively correlated. Outcomes close to the origin of the correspondence plot correspond to low correlations. A somewhat puzzling phenomenon is the isolated location of "disagree strongly" in the correspondence plot. The balloon plot indicates that this outlying position results from the fact that for France this category is highly over-represented, while for all other countries this response category behaves quite similar.

It may be interesting to see how a biplot for the correlation looks like. To this end I apply an SVD to the correlation matrix $R = (\varrho_{k\ell})$ resulting in $R = UDV'$. Let U_2 (V_2) denote the first two columns of the matrix of the left (right) singular vectors and D_2 is the upper-left (2 × 2) submatrix of D. Figure 3 presents the (symmetric) biplot based on $U_2 D_2^{1/2}$ (as row coordinates) and $V_2 D_2^{1/2}$ (as column coordinates). Overall the biplot resembles the original correspondence plot in Figure 2 but some interesting differences emerge. In Figure 3 the category "disagree strongly" is no longer as isolated as in the original correspondence plot. Now this category is located much closer to "France" (such that their labels overlap) representing the high correlation between "France" and "disagree strongly". Furthermore, the biplot also reveals the outlying position of the combination

44

"strongly agree" and "Russia". This corresponds to the largest correlation of the whole table with nearly 0.295, whereas the correlation between "France" and "disagree strongly" is substantially lower (0.18).

7. Conclusion

As it's time to leave the planet "correspondence analysis" I am going to leave a message in the bottle at the Schwarzrheindorfer waterfront: "Whoever finds this bottle, let it be said that 65 years after the fateful 1957 an alien from a distant planet has strayed into this inhospitable territory. He left some cryptical notes on alternative distance measures, although everyone on this planet is happy about this geometry, which is not encountered anywhere else." So in this bottle you will find a collection of red and green balloons. If you want to escape the limitations of a two-dimensional plane, you may fill these balloons with helium and they will carry you away, right in the direction of my home planet...".

References

Bai, J. (2003): Inferential Theory for Factor Models of Large Dimensions. In: Econometrica 71, pp. 135–172.
Blasius, Jörg (2001): Korrespondenzanalyse. München: Oldenbourg.
Blasius, Jörg/Greenacre, Michael (2006): Correspondence Analysis and Related Methods in Practice. In: Greenacre, Michael and Blasius, Jörg (eds.): Multiple Correspondence Analysis and Related Methods. London: Chapman & Hall/CRC. pp. 3–40.
Breitung, Jörg (2013): Factor Models. In: Hashimzade, Nigar and A. Thornton, Michael A. (eds.): Empirical Macroeconomics. Edward Elgar, pp. 249–265.
Breitung, Jörg/Eickmeier, Sandra (2011): Testing for structural breaks in dynamic factor models. In: Journal of Econometrics, 163, pp. 71–84.
Breitung, Jörg/Tenhofen, Jörn (2011): GLS estimation of dynamic factor models. In: Journal of the American Statistical Association, 106, pp. 1150– 1166.
Greenacre, Michael (2007): Correspondence Analysis in Praxis. 2nd ed. London: Chapman & Hall/CRC.
Nenadić, Oleg/Greenacre, Michael (2007): Correspondence Analysis in R, with Two- and Three-dimensional Graphics: The ca Package. In: Journal of Statistical Software, Volume 20, Issue 3.
Rao, C.R. (1995): A review of canonical coordinates and an alternative to correspondence analysis using Hellinger distance. Qüestiió, 19, pp. 23–63.
Yang, Wei/Xu, Luhui/Chen, Xiaopan/Zheng, Fengbin/Liu, Yang (2015): Chi-Squared Distance Metric Learning for Histogram Data. In: Mathematical Problems in Engineering, 2015, Volume 2015, Article ID 352849.

Applying Combinatorial Inference in GDA. The Case of European Central Bank Governing Council Members (1999–2022)

Frédéric Lebaron, Brigitte Le Roux, Aykiz Dogan

Abstract

Based on previous studies by the authors, this contribution proposes an integrated methodology applied to an economic sociological example which is the leading decision-making elites of the European Central Bank (ECB). This methodology responds to the limits of quantitative analysis on small populations such as elite decision-makers focusing on the case of leading monetary policy-makers which became the object of a growing literature. The methodological approach explained in this chapter uses prosopography and Geometric Data Analysis (GDA) to explore the particular case of a small but exhaustive data set which summarises biographical characteristics of ECB Governing Council members since its establishment in 1999 (n=85). This data is crossed with position-takings summarised in terms of hawk, dove and intermediary monetary policy orientations to propose statistical interpretations of the determinants of economic decisions. The article details the basic steps of a multiple correspondence analysis (MCA). This procedure first involves the construction of a social space on the basis of indicators of various species of capital. We then study a structuring factor of interest (which is, in this case, the individual's monetary orientation or "position-taking" and secondly gender) in the cloud of individuals to assess its descriptive effects. Finally, we apply combinatorial inference to study the atypicality and the compatibility zone around the mean point of a particular sub-group, and the heterogeneity between two groups. In this way, we present a methodological approach to explore the factors determining position-takings within the ECB Governing Council. We argue that this methodological approach is particularly helpful in dealing with very low frequencies and relatively scarce or small data, without leaving aside the multidimensionality and complexity of the object. This chapter hence contributes both to methodological discussions on the study of small groups of elite leaders and to empirical studies on the ECB and more broadly on central banks' policies by providing a sociological statistical analysis of policy-makers' properties that influence policy orientations and outcomes.

1. Introduction

A recurring difficulty in the study of small elite groups by statistical methods is the small number of cases which constrains quantitative analysis. At the top positions in the global social space, only a few individuals occupy positions of power, which allow them to produce and to impose a legitimate world vision (Bourdieu 1989), to make highly influential decisions having major consequences (Mills 2019). The study of these power positions requires identifying the mechanisms which allow access to only certain actors with specific characteristics, and hence the resources and capitals which distinguish these actors. Therefore, descriptive biographies of actors including comparisons over time and space have been a central methodological tool in the study of the "field of power" (Bourdieu 1989: 16), often using "prosopography" (see for example, Bourdieu 1989, Charle 2013, Dogan and Lebaron 2023). Recent scholarship on elites, especially in the case of decision-making leaders, has furthermore put emphasis on the systematic ways that personal attributes and life experiences of individual leaders affect political outcomes (Krcmaric, Nelson, and Roberts 2020).

The difficulty of studying very small but strongly influential populations is highlighted by Lebaron and Dogan (2020) in the case of central bank governors and committee members, which concentrate an important monetary and financial regulatory power impacting millions of citizens in their daily life. Recent responses of central banks to the financial crisis of 2008 and to the effects of the Covid-19 pandemics made it clear that this impact is not limited to the value of the currency or the price levels but extends on the whole economic social relations with large scale consequences on overall economic conditions (e.g. Warjiyo and Juhro 2022) and development (Epstein 2019) as well as socioeconomic inequalities and class interests (see for example Canova 2015; Dietsch, Claveau, and Fontan 2018; Seccareccia 2017; Young 2018). In fact, monetary decisions have always been a major component of economic policy, which might be defined as the way public actors try to define, monitor and regulate economic dynamics that are, to a great extent, "private" in capitalist economies.

A large corpus of economics and economic history literature has investigated the various mechanisms of monetary decision, leading to strong doctrinal and theoretical controversies, and to a large accumulation of empirical data on the formation of interest rates or other key components of monetary policy. Recently this literature started to pay greater attention to individual decision-makers. Various studies showed that central bank governors actively contribute in legitimising monetary policy and institutions while at the same time reproducing particular social values and norms (Lebaron 2000) and collective identities including national (Sørensen 2015) and gender identities (Clarke and Roberts 2016). In parallel, the idea has gained ground that central bank leaders' background has an impact on their policy choices and monetary reasoning.

An increasing number of studies have investigated biographical characteristics which influence these decision makers' policy choices and position takings including educational and occupational background (Göhlmann and Vaubel 2007), gender (Diouf and Pépin 2017), political affiliation (Neuenkirch and Neumeier 2015), religious orientations (Lebaron and Dogan 2023), doctrinal or theoretical beliefs (Romer and Romer 2004) or personal experience of particular macroeconomic conditions (Malmendier, Nagel, and Yan 2021). A small number of studies have explored multiple variables (e.g. Chappell, McGregor, and Vermilyea 2004; Farvaque, Stanek, and Hammadou 2011). Notably, Lebaron and Dogan (2016, 2018, 2020) have highlighted that these multivariable characteristics and identities combine and operate in very complex multidimensional ways which cannot be grasped through simplistic determinist approaches.

Overall, this growing literature supports the hypothesis that biographical characteristics of governors and committee members are important in understanding the way central banks "behave", that is the way they make and implement decisions. This hypothesis is consistent with previous findings and arguments of Bourdieu (2022) suggesting that position-takings in the field of power directly relate to positions and trajectories in this field and operate through the complex machinery of habitus. Putting individuals at the centre of the analysis derives from this conception of action where practices of agents depend not only on immediate calculation but on a set of internalised dispositions which lead them to particular views and preferences. These relations need to be described and statistically assessed in each particular decision-making process. But is there a way to provide systematic interpretations on the *effects* of these biographical characteristics in the case of small committees while taking account of their complex multidimensional relationship?

In this contribution, we develop a methodological approach for studying very limited though powerful populations. This approach is based on the application of Multiple Correspondence Analysis (MCA) as a descriptive method for examining relationships among several categorical variables (Le Roux and Rouanet 2010), such as socio-demographic, biographical and behavioural (position-takings) properties of a set of individuals. This method is widely used in social sciences since the exemplary work of Bourdieu who combined it with other Geometric Data Analysis (GDA) techniques to study field structures (Lebaron and Le Roux 2018). This method is particularly useful for establishing a "spatial" representation of the positions of field actors according to the "distance" between them in terms of shared characteristics or differences, hence to present a visual summary of the relationship between selected variables. This method allows us to deal with low frequencies (due to small number of individuals) and small data, without leaving aside the multidimensionality and complexity of the object.

As a case-study, we propose to explore the factors which influence position-takings of the European Central Bank (ECB) Governing Council members since 1999. During this period, the ECB has had to deal with the construction of a coherent enlarging Eurozone, the establishment of budgetary rules consistent with its mission, the financial crisis of 2007–2009, sovereign debt crisis after 2010, the generalisation of quantitative easing measures until 2019, and more recently the shocks of the pandemic and the effects of the come-back of inflation due to the energy crisis. This has been a rich and highly challenging period for this new supranational monetary institution. Our study will provide an analysis of the actors behind the scenes in this process: the members of the Governing Council (GC) which meets twice a month and decides various components of the public policy of the institution adopting guidelines for member states, piloting monetary policies in the Eurozone, defining monetary objectives, key interest rates, the supply of reserves, the framework of banking supervision, and so on[1]. The GC consists of six Executive Board (EB) members and the national central bank governors of the Eurozone which currently includes twenty member states.

Since the establishment of the ECB in 1998, 85 individuals have worked in this Council. Even though this number allows reliable statistical calculations, the problem of working with small populations is particularly evident in sub groups. A significant example results from the gender disparity contested by the European parliament on several occasions[2]. The 7 women who participated in the GC since its establishment represent a proportion so low (0.08) that statistical methods might produce biased results in assessing the effect of gender on position-takings. Within this framework, the study of gender dimension is of particular interest for the methodological approach we present here for the study of small elite populations as well as our sociological analysis of common

---

1 ECB Governing Council. https://www.ecb.europa.eu/ecb/orga/decisions/govc/html/index.en.html
2 The European Parliament's concerns are clearly manifested in its resolutions. For example, the 2020 annual report of texts adopted contests that "whereas, notwithstanding the repeated calls of the European Parliament to receive a gender-balanced shortlist of at least two names for ECB Executive Board positions, the shortlist for the appointment of a new member [...] was composed only of men; whereas women continue to be strongly under-represented on the Governing Council". The related resolution "Expresses strong concern that only two of the 25 Members of the ECB's Governing Council are women, despite repeated calls from Parliament and from senior figures in the ECB, including its President Christine Lagarde, to improve gender balance in EU economic and monetary affairs nominations". These concerns are repeated in the 2021 annual report of texts adopted as well. European Parliament resolution of 10 February 2021 on the European Central Bank – annual report 2020. https://www.europarl.europa.eu/doceo/document/TA-9-2021-0039_EN.html For a more recent dataset: https://www.econostream-media.com/news/2023-02-01/ecb_hawk-dove_ranking:_lagarde_lane_and_other_governing_council_members_reranked.html

and particular characteristics of monetary actors and the complex intersectional relationships between these characteristics.

The position-takings of the GC members are grouped according to three general categories: hawk / dove / intermediary or uncertain stances. This is a common and largely diffused categorization in the field of central banking. Dovishness refers to positions favourable to a proactive use of monetary policy such as promoting full employment whereas hawkishness summarises a high level of sensitivity to inflation and importance attributed to price stability as a primary policy goal. Hawks tend to favour higher interest rates and restrictive measures whereas doves on the contrary favour relatively flexible interest rates and accommodative measures (such as, in our period, quantitative easing measures).

In the following sections, we present different stages of our methodological approach. We first present prosopographical data collection as a primary step. We introduce the principles of what Bourdieu (1985) called the "construction of the social space" to identify individual positions in a field based on the distribution of different varieties of capital. Then we provide a descriptive analysis of the results of the MCA applied to data on GC members. We study a structuring factor (which is, in this case, the individual's monetary orientation or "position-taking" and secondly gender) in the cloud of individuals to assess descriptive effects. Finally, we apply combinatorial inference to study the atypicality of a particular sub-group, the compatibility zone around the mean point, and the heterogeneity between two groups.

2. Data collection

Our methodology is based on the "prosopographical" approach to data collection explained by Dogan and Lebaron (2023). For this study, we constructed a prosopographical database by collecting biographical information on each individual of the study population (GC members). The principal sources of data were CVs or short biographies found in the websites of the ECB and national central banks. We also consulted external biographical sources such as national Who's who, Who's who in Central Banking, Wikipedia, LinkedIn, inter/national press and other online sources to complement and crosscheck information. We coded education separately for the degree (PhD, Master and other), the main discipline and the location of study abroad if any. The professional experience is coded for seven main sectors (see the following section): it can be the main experience of the individual, it can be a "secondary" experience, or the individual has no experience in the sector. This coding technique has the advantage of including standardised yet nuanced information.

For coding the position-takings, we have consulted various online institutional sources, press declarations as well as hawk/dove analysis provided by "central bank watchers". More specifically, we used rankings by Business Insiders and EconoStream Media which provide a hawk/dove score based on numerical scales (see figure 1). Based on these scores, we coded the position-takings of each member on a three-categories scale: Dove / Intermediary / Hawk.

Figure 1. ECB Hawk/Dove Ranking.

ECB - Hawk/Dove Ranking
Last Update: 17 February 2023 — Econostream Media

Name	Position	Hawk Scale	16-Mar	04-May	15-Jun
Most Dovish					
Panetta	Exec Board		✓	✓	✓
Stournaras	Greece		✗	✓	✓
Visco	Italy		✓	✓	✓
Lane	Exec Board		✓	✓	✓
Hernández de Cos	Spain		✓	✗	✓
Herodotou	Cyprus		✗	✓	✓
Centeno	Portugal		✓	✓	✗
Makhlouf	Ireland		✓	✓	✓
de Guindos	Vice President		✓	✓	✓
Lagarde	President		✓	✓	✓
Scicluna	Malta		✓	✗	✗
Villeroy	France		✓	✓	✗
Rehn	Finland		✓	✓	✓
Reinesch	Luxembourg		✓	✗	✓
Vujčić	Croatia		✗	✓	✓
Schnabel	Exec Board		✓	✓	✓
Elderson	Exec Board		✓	✓	✓
Vasle	Slovenia		✓	✓	✗
Šimkus	Lithuania		✓	✗	✓
Kazāks	Latvia		✗	✓	✓
Kažimír	Slovakia		✓	✓	✓
Müller	Estonia		✓	✓	✓
Wunsch	Belgium		✓	✓	✓
Nagel	Germany		✓	✓	✓
Holzmann	Austria		✓	✗	✗
Knot	Netherlands		✗	✓	✓
Most Hawkish					

Source: Econostream Media, March 2023. This source is regularly up-dated: https://www.econostream-media.com/hawk-dove.html

It should be noted that before 2015, each governor had one voting right. Since their number exceeded 18 as of 1 January 2015 when Lithuania joined the euro area, country governors use their voting right in monthly rotations. Hence contrary to the EB members, country governors' voting rights vary in the rotation system which, implemented since 2015, classifies member countries into two groups according to the "size of their economies and their financial sectors" and attributes weighted rights to them[3]. This system thus distorts power in favour of larger economies.

3. Construction of a social space

The first step of the analysis is the visualisation of the social space of the ECB's decision-makers on the basis of collected information. Following Bourdieu's precursory work on the study of social fields, we construct a social space on the basis of different kinds of power or "capital"[4] which are both empirically accessible and sociologically meaningful to identify patterns of relationships and distances between positions in the "social topology" (Bourdieu 1985)[5]. First, we examine educational trajectories and background, professional experience in different sectors, and other positional institutional variables which might be influential. We follow here a set of empirical studies by Lebaron (2008), Lebaron and Dogan (2016, 2018), that have documented the existence of a "social space of the Governing council", which regroups the leading monetary decision-makers of the Eurozone since its establishment.

In our case-study, the data summarises three key elements. First, each individual's position in the field is defined based on two indicators: 1) status in the GC (past or present EB member or country governor); 2) country group (instead of using the two groups of the rotation system, we identify poles based on the relationship of nationality with other variables). The second active property is educational background defined by three indicators: 1) level of education; 2)

---

[3] "The Governors from countries ranked first to fifth – currently, Germany, France, Italy, Spain and the Netherlands – share four voting rights. All others (15 since Croatia joined on 1 January 2023) share 11 voting rights. The Governors take turns using the rights on a monthly rotation." Rotation of voting rights in the Governing Council. (Last update on 1 January 2023) https://www.ecb.europa.eu/ecb/educational/explainers/tell-me-more/html/voting-rotation.en.html

[4] "The kinds of capital, like the aces in a game of cards, are powers that define the chances of profit in a given field" Bourdieu (1985: 724).

[5] According to Bourdieu (1985: 723–4) "the social world can be represented as a space (with several dimensions) constructed on the basis of principles of differentiation or distribution constituted by the set of properties active within the social universe in question, i.e., capable of conferring strength, power within that universe, on their holder. Agents and groups of agents are thus defined by their relative positions within that space".

main discipline; 3) location of studies abroad. The third element is the professional trajectory defined by seven indicators as binary answers (yes/no) to work experience in 1) administration; 2) international organisations; 3) university; 4) politics; 5) central bank; 6) finance; 7) private sector. We proceed to a MCA of 12 active variables (see Table 3 in the Appendix).

4. Interpretation of the axes

The interpretation will be based on the first three axes which form a three dimensional space. Their variances are λ1=0.228, λ2=0.182, λ3=0.173 (that of Axis 4 is 0.145). The cumulated modified rates of the first 5 axes reaches 83%.

Figure 2: Scree plot: variance of axes (eigenvalues); cumulated variance rates and cumulated modified rates

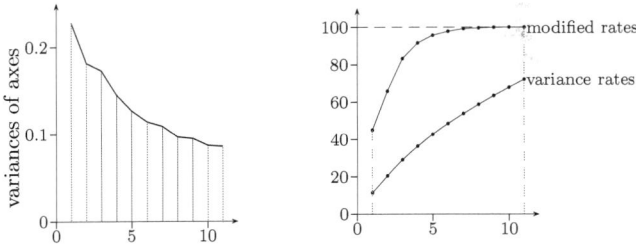

The axes are interpreted by using the method of contributions: we use the cloud of categories. As a general rule, the interpretation of an axis is based on the categories whose contributions exceed the average contribution (100/36 = 2.8%).

Figure 3 shows the 15 categories contributing most to the first axis. Together they contribute to 79 percent of the variance of axis 1 (see appendix, Table 4).

Figure 3: Interpretation of axis 1: 15 categories most contributing to axis 1

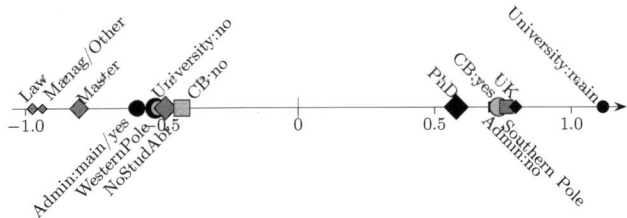

Axis 1 describes an opposition between, on one pole, law or management graduates (mostly master level) who previously worked in public administration and, on the other, academics, mostly from Southern Europe, with PhD, studies abroad (in particular UK) and previous experience in central banking (PhD-main career university-Southern pole-studies in the UK *versus* Law-Management-Administration). *This is an academic versus bureaucratic capital axis.*

Figure 4 shows the 13 categories contributing most to the second axis. Together they contribute to 82 per cent of the variance of axis 2 (see appendix, Table 5).

Figure 4: Interpretation of axis 2: 13 categories most contributing to axis 2

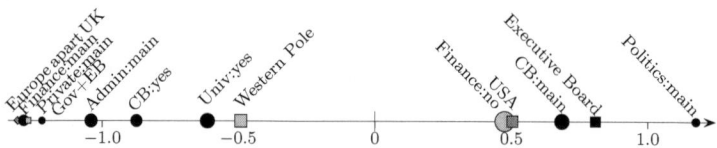

Axis 2 expresses an opposition between politicians or central bankers who join the GC as EB members and those who built their career in private or public finance and had previous experience in central banking or academics (Political-Executive Board-Internal-US education Versus Academy-Administration-Finance). *It is a political and insider capital versus finance capital axis.* While the first group's studies abroad were mainly in the US, the second group's were in Europe.

Figure 5 shows the 11 categories contributing most to the third axis. Together they contribute to 74 percent of the variance of axis 3 (see appendix, Table 6).

Figure 5: Interpretation of axis 3: 11 categories most contributing to axis 3.

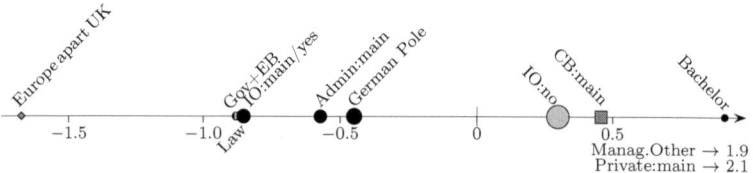

Axis 3 opposes law graduates and careers in international organisations to careers in private sector or central banking (International Organizations-Europe-Gov&Executive Board-Law Versus Internal-Bachelor-Private). On the left side, IO careers and law studies relate to studies abroad in European universities and seem to provide access to the GC as EB members. On the right side, private careers are associated with lower degrees (bachelor) in management sciences. *This is an axis of international and juridical capital versus private or insider capital.*

The space of members of the Governing Council constructed through this MCA is clearly multidimensional and one finds the classical sources of differentiation in the field, namely academic versus bureaucratic capital (axis 1), political and insider capital versus *finance capital* (axis 2) and finally international and legal capital versus private and internal careers (axis 3).

5. "Position taking" as a structuring factor

A *structuring factor* generates sub-clouds of individuals. We begin by studying the "position taking" as a structuring factor.

The following table gives the coordinates of the mean points of the sub-clouds "hawk" and "dove" as well as their scaled deviation (see Appendix A1).

Table 1: Coordinates of the two mean points of the sub-clouds Dove and Hawk on the first three principal axes and plane 1–3 and scaled deviations

	size	coordinates			scaled deviations from origin			
		axis 1	axis 2	axis 3	axis 1	axis 2	axis 3	plane 1-3
dove	24	0.173	0.010	0.111	0.36	0.02	0.27	0.45
hawk	25	−0.110	0.021	−0.074	−0.23	0.05	−0.18	0.29

Descriptively, we see that

1. for "dove", the scaled deviation from the origin is above the 0.4 threshold in plane 1–3.
2. the position takings "dove" and "hawk" appear to be different in plane 1–3

Figure 6: Cloud of individuals in plane 1–3 with the mean points of the dove (black circles) and hawk sub-clouds (gray circles) and the compatibility zone for dove group.

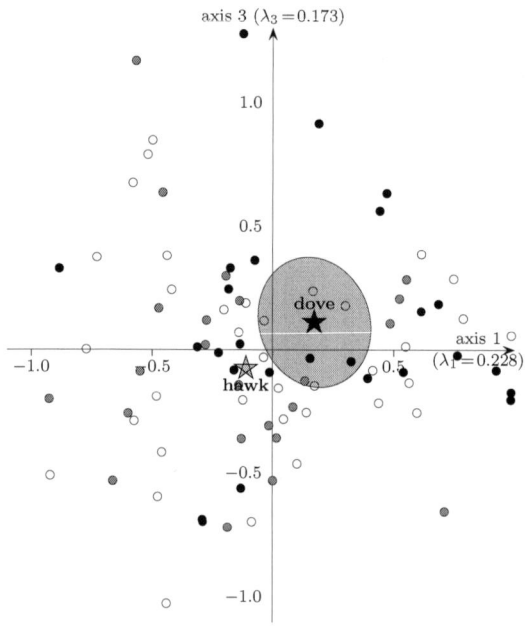

Following the descriptive conclusions, we proceed to inductive analysis by studying the level of typicality of the group "dove" and the homogeneity of the two subgroups "dove" and "hawk" in plane 1–3. For this purpose, we will use the combinatorial inference method based on permutation tests[6].

6. Typicality of "Dove" group

In order to assess the level of typicality of "dove" responses within the set of ECB GC members, we will use the combinatorial typicality test (see Appendix A2.). For the "dove" group, the number of all possible samples is more than 9×10^{20}. This number is very large, hence the exact test of typicality is carried out through the Monte Carlo method by drawing one million random samples.

The *combinatorial distribution* of the test is shown in Figure 7.

---

6   Combinatorial inference is in harmony with the key ideas of Geometric Data Analysis: "description comes first" and it is provided in a multidimensional setting. Effects are investigated provided that they are sufficiently large.

Figure 7: Typicality of dove; Permutation distribution of the test statistic (based on 10^6 samples), the observed value is 0.20, the p-value is 0.033.

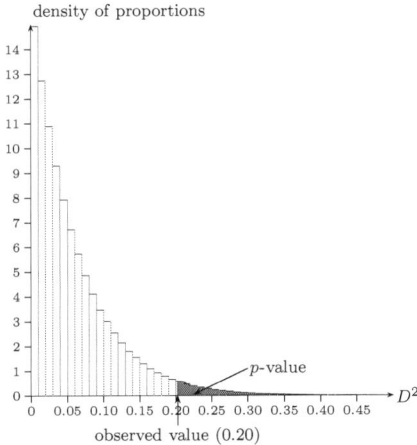

For the "dove" group, the p-value is equal to 0.033 <0.05. We conclude that, in the plane 1–3, the "dove group" is atypical of the overall set of ECB GC members with respect to the *mean point*, at level 0.05.

We now treat the problem of assessment of location of the mean point of the "dove" sub-cloud by using the geometric typicality test (see Appendix 2.2.). The compatibility region for the mean point of the "dove" sub-cloud is mainly in the upper-right quadrant (see Figure 6 above).

A conclusion is that, according to the mean point, the group "dove" is atypical in plane 1–3, on the side of academic capital and towards the "private" pole. This result is consistent with previous findings that relate dovish positions to more academic and less bureaucratic trajectories in this particular field (Lebaron, Dogan, 2016, Le Roux, Bienaise, Durand, 2019). This relation is based on a particular dispositional effect: academics in Europe tend to be less committed to the neoliberal doctrinal conceptions of monetary policy which have spread in the bureaucratic field both in Germany (Bundesbank, ministry of finance) and in French financial administration. We might assume that they are more open to international academic debates, which confront a diversity of perspectives including unorthodox or critical approaches to neoliberal doctrines. This might explain that academics have a broader theoretical perspective which allows them to experiment with less orthodox monetary policies and instruments such as those associated with "welfare states" or "economic states", or newer policies developed as responses to recent crises.

We observe that the mean point of the "hawk" sub-cloud is not in the compatibility region. So we will study more precisely the heterogeneity of the groups "dove" and "hawk".

7. Homogeneity of "dove" and "hawk" groups

We will now assess the level of homogeneity of the two groups, and thus try to corroborate the descriptive conclusion of difference between the two groups, in plane 1–3. To do this we use the homogeneity test (see Appendix A2.3.). In this particular case, we take into account the cloud of all ECB GC members, then we proceed to a partial comparison between the two groups. The histogram of the permutation distribution of the test statistic is shown in Figure 8.

Figure 8: Homogeneity "dove-hawk": Permutation distribution of the test statistic (homogeneity test) (based on 10^6 samples), the observed value is 0.549, the p-value is 0.034.

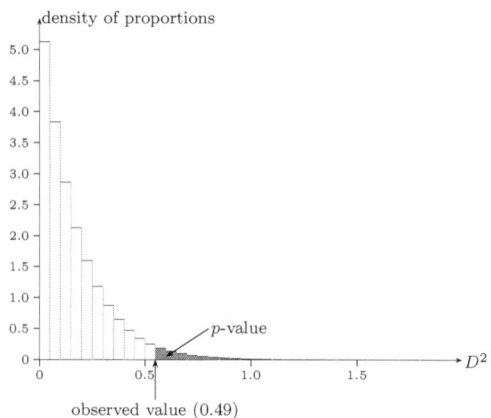

Then, the homogeneity level of the two sub-clouds with respect to the overall cloud of ECB GC members is defined by the proportion of possible pairs of sub-clouds (that are comprised of 24 and 25 points) whose statistic value is greater than (or equal to) that of the observed pair ("dove", "hawk"). This proportion (p-value) is equal to 0.034 < 0.05. Hence, we can conclude that in plane 1–3, "dove" and "hawk" groups are heterogeneous, at level 0.05.

8. Sociological comments

Axes 1 and 3 combined illustrate the importance of academic and private capital in this field, on the side of dovishness. This is clearly related to the general configuration of the Governing Council where academic and private trajectories are relatively numerous, but not the most strongly related to orthodox monetary positions, which are related to bureaucratic careers in dominant institutions such as the Bundesbank, the Banque de France or the ministries of finance of bigger countries.

9. Study of a very small group: women

We study here the specificity of the 7 women in the cloud of individuals, in the same constructed space.

Figure 9: Cloud of individuals in plane 1–2 with the sub-cloud of women (black points) and the mean points of the dove and hawk sub-clouds (black star and grey star).

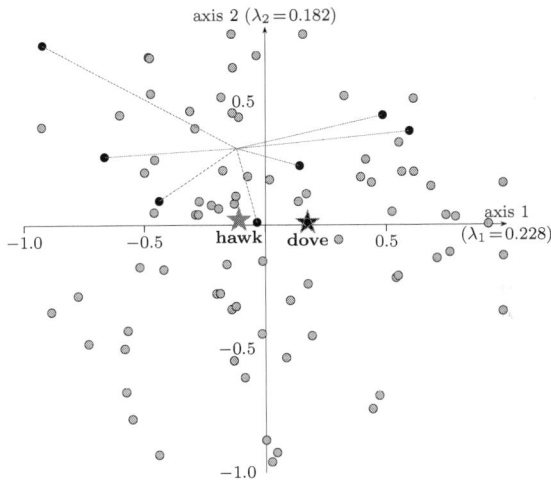

This group is of particular sociological interest for understanding the way social dimensions, here gender, participate in systematic differences observed inside the committees. Whereas it can be argued that women are hired first because they conform to the doctrinal monetary norm (which is partially the case in Europe), one can investigate at least their specificity in terms of trajectories. In this line of research, women are characterised by particular paths inside the field of power.

Table 2: Coordinates of the mean points of the two sub-clouds woman and man on the first three principal axes and scaled deviations

	size	coordinates			scaled deviations from origin		
		axis 1	axis 2	axis 3	axis 1	axis 2	axis 3
woman	7	−0.119	0.310	−0.135	−0.250	0.729	−0.325
man	78	0.011	−0.028	0.012	0.022	−0.065	0.029

The scaled deviation between the mean point of the woman sub-cloud and the origin is large (above the 0.4 threshold) on axis 2. But there are only 7 women in the group, so we will perform the typicality test (for axis 2) to be sure that the result is significant, that is, not due to chance. The combinatorial *p*-value is equal to 0.019: on the second axis the woman group is atypical of the overall cloud, on the side *"political and insider capital"*. This is consistent with historical information pointing to the importance of international capital for women, as well as political capital in the context of the Eurozone (Lebaron and Dogan, in preparation).

As their number is very low, this is a rather interesting result, which should of course be tested in other contexts, in order to assess the specificity of the Eurozone context or the extension of the phenomenon. By construction, it is equivalent to say that women are atypical of the 85 ECB group or that they are heterogeneous to men on axis 2.

Five women have intermediary or unknown position-takings. On the basis of combinatorial inference, it is impossible to conclude to their atypicality on one or several of the axes, as well as their heterogeneity with the other members of our population. Here again, the limitation comes from the small size of the sub-population which does not allow to discard the effects of random factors or chance.

10. Conclusion

Our approach has allowed us to reach inferential conclusions about effects detected in limited populations, which go beyond purely descriptive conclusions and substantiate our biographical approach. *Dovish* positions relate to particular types of biographical trajectories in the field, which can be characterised by more academic and more "private" careers. Past and present female members of the Governing Council are most likely to gain access to this position after international trajectories and the accumulation of political capital relevant at the EU level (which is strongly related to the appointments in the Executive Board). Not only the small number of women, but also their intermediary stances between hawk and dove orientations limit the possibility to obtain inferential conclusions and discard the effect of chance.

The methodological approach defended in this chapter is strongly based on a multidimensional and geometric conception of social reality, substantiated by Geometric Data Analysis. It helps to describe social spaces by configurations of properties defining specific forms of capital, and to investigate the effects of positions in those spaces or *field effects* both descriptively and inferentially. As inferential procedures are based on the combinatorial inference framework, our analysis is in coherence with the multidimensional mapping of data and the emphasis put on social actors.

From a sociological point of view, we have been able to assess and visualise effects on very small populations: position-takings relate to certain biographical characteristics, which must be further examined through case-studies. A more qualitative sociological interpretation becomes possible and can be substantiated by a statistical multidimensional basis.

References

Bourdieu, Pierre (1985): The social space and the genesis of groups. In: Social Science Information, 24, 2, pp. 195–220.
Bourdieu, Pierre (1989): Social space and symbolic power. In: Sociological Theory, 7, 1, pp. 14–25.
Bourdieu, Pierre (2000) : Les structures sociales de l'économie. Paris: Seuil.
Bourdieu, Pierre (2022): Microcosmes. Théorie des Champs. Paris: Raisons d'agir.
Canova, Timothy A. (2015): The role of central banks in global austerity. In: Ind. J. Global Legal Stud., 22, p. 665.
Chappell Jr., Henry W./McGregor, Rob Roy/Vermilyea, Todd (2004): Committee decisions on monetary policy: Evidence from historical records of the Federal Open Market Committee. MIT press.
Charle, Christophe (2013): La prosopographie ou biographie collective. Bilan et perspectives. In: Charle, Christophe (ed.): Homo historicus. Réflexions sur l'histoire, les historiens et les sciences sociales. Paris: Armand Colin, pp. 98–102.
Clarke, Chris/Roberts, Adrienne (2016): Mark Carney and the gendered political economy of British central banking. In: The British Journal of Politics and International Relations, 18, 1, pp. 49–71.
Dietsch, Peter/Claveau, François/Fontan, Clément (2018): Do central banks serve the people?. John Wiley & Sons.
Diouf, Ibrahima/Pépin, Dominique (2017): Gender and central banking. In: Economic Modelling, 61, pp. 193–206.
Dogan, Aykiz/Lebaron, Frédéric (2023 forthcoming): "Prosopography". In: Badache, Fanny/Kimber, Leah R./Maertens, Lucile (eds.): International Organizations and Research Methods. Michigan University Press.

Epstein, Gerald (2019): Central Banks as Agents of Economic Development. In: The Political Economy of Central Banking. Edward Elgar Publishing, pp. 407–425.

Farvaque, Etienne/Stanek, Piotr/Hammadou, Hakim (2011): Selecting your inflation targeters: Background and performance of monetary policy committee members. In: German Economic Review, 12, 2, pp. 223–238.

Fisher, Ronald A. (1935): The Design of Experiments. Edinburgh: Oliver & Boyd.

Fontan, Clément (2018): Frankfurt's double standard: the politics of the European Central Bank during the Eurozone crisis. In: Cambridge Review of International Affairs, 31, 2, pp. 162–182.

Göhlmann, Silja/Vaubel, Roland (2007): The educational and occupational background of central bankers and its effect on inflation: An empirical analysis. In: European Economic Review, 51, 4, pp. 925–941.

Krcmaric, Daniel/Nelson, Stephen C./Roberts, Andrew (2020): Studying leaders and elites: The personal biography approach. In: Annual Review of Political Science, 23, pp. 133–151.

Le Roux, Brigitte/Rouanet, Henry (2010): Multiple correspondence analysis. Vol. 163. Sage.

Le Roux, Brigitte/Bienaise, Solène/Durand, Jean-Luc (2019): Combinatorial Inference in Geometric Data Analysis, Boca Raton – London – New York: Chapman & Hall/CRC.

Le Roux, Brigitte/Rouanet, Henry (2004): Geometric Data Analysis: From Correspondence Analysis to Structured Data Analysis (Foreword by P. Suppes). Dordrecht: Kluwer.

Lebaron, Frédéric (2008): Central bankers in the contemporary global field of power: a 'social space' approach. In: The Sociological Review, 56, 1_suppl, pp. 121–144.

Lebaron, Frédéric (2000): La croyance économique. Les économistes entre science et politique, Paris: Seuil.

Lebaron, Frédéric/Dogan, Aykiz (2020): Central bankers as a Sociological Object. Stakes, Problems and Possible Solutions. In: Denord, Francois/Palme, Mikael/Réau, Bertrand (eds.): Researching Elites and Power: Theory, Methods, Analyses. Springer, pp. 95–112.

Lebaron, Frédéric/Dogan, Aykiz (2023 forthcoming): L'ethos du central banker. La religion comme capital dans le champ global des banques centrales. In: Galonnier, Juliette/Fer, Yannick/Angey, Gabrielle (eds.): Sociologie critique des religions. Paris, ENS Editions.

Lebaron, Frédéric/Dogan, Aykiz (2018): Social Dimensions of Monetary Decisions. A Study of the ECB Governing Council. In: Champroux, Nathalie/Depeyrot, Georges/Dogan, Aykiz/Nautz, Jürgen (eds.): Construction and Deconstruction of Monetary Unions, Lessons from the Past. Wetteren, Moneta 201.

Lebaron, Frédéric/Dogan, Aykiz (2016): "Do central banker's biographies matter?" In: Sociologica 2.

Lebaron, Frédéric/Dogan, Aykiz (in preparation): "Feminization of Central Banks. A socio-historical study on female central bank governors". Working Paper.

Lebaron, Frédéric/Le Roux, Brigitte (2018): "Bourdieu and Geometric Data Analysis." In: Medvetz, Thomas/Sallaz, Jeffrey J. (eds.): The Oxford Handbook of Pierre Bourdieu. Oxford University Press.

Malmendier, Ulrike/Nagel, Stefan/Yan, Zhen (2021): The making of hawks and doves. In: Journal of Monetary Economics, 117, pp. 19–42.

Mills, Charles Wright (2019): The power elite. In: Social Stratification. Routledge, pp. 202–211.

Neuenkirch, Matthias/Neumeier, Florian (2015): Party affiliation rather than former occupation: The background of central bank governors and its effect on monetary policy. In: Applied Economics Letters, 22(17), pp. 1424–1429.

Pitman, Edwin J. (1937): Significance tests which may be applied to samples from any population. In: Journal of the Royal Statistical Society, 4, pp. 119–130.

Romer, Christina D./Romer, David H. (2004): Choosing the Federal Reserve chair: lessons from history. In: Journal of Economic Perspectives 18, 1, pp. 129–162.

Schmidt-Wellenburg, Christian/Bernhard, Stefan (eds.) (2020): Charting Transnational Fields. Methodology for a Political Sociology of Knowledge. New York: Routledge.

Seccareccia, Mario (2017): Which vested interests do central banks really serve? Understanding central bank policy since the global financial crisis. In: Journal of Economic Issues, 51, 2, pp. 341–350.

Sørensen, Anders Ravn (2015): Banking on the nation: How four Danish central bank governors used and reproduced the logics of national identity. In: International Journal of Politics, Culture, and Society, 28, pp. 325–347.

Warjiyo, Perry/Juhro, Solikin M. (eds.) (2022): Central Bank Policy Mix: Issues, Challenges, and Policy Responses. Handbook of Central Banking Studies. Springer Nature.

Young, Brigitte (2018): Financialization, unconventional monetary policy and gender inequality. In: Elias, Juanita/Roberts, Adrienne (eds.): Handbook on the International Political Economy of Gender. Edward Elgar Publishing, pp. 241–251.

Appendix

A1: Note about scaled deviations

If y_1 denotes the principal coordinate of the mean point of a subcloud on axis 1, the scaled deviation from the mean point to the origin on axis 1 (whose variance is λ_1) is equal to $d_1 = y_1/\sqrt{\lambda_1}$. Similarly, the scaled deviation on axis 2 is $d_2 = y_2/\sqrt{\lambda_2}$. The scaled deviation in plane 1–2 is equal to $\sqrt{(d_1^2 + d_2^2)}$ (see Le Roux 2019, p. 22). A scaled deviation is regarded as notable if it is greater than 0.4 and small if it is less than 0.3, but these values should reconsidered in each practical situation.

A2: Combinatorial inference

In this chapter, we outline statistical inference procedures for GDA that are based on a combinatorial framework and that highlight the role of *permutation tests*[7]. The methods relate mainly to studying a *Euclidean cloud*. Permutation tests belong to the set of resampling methods, so called because the data are resampled and reexamined repeatedly in order to obtain results. They are *data-dependent*: all the information requested for analysis is contained within the observed data. No assumptions regarding the distribution of the data are required.

A2.1 Combinatorial typicality test

In order to offer a solution to the typicality problem, the basic idea is to compare the group to the samples of the reference set, where samples are simply defined as subsets of the reference set. For this purpose, we construct the set of all possible samples that have the same number of elements as the group, and we locate the observations of the group among the ones of the possible samples according to some statistic of interest (test statistic). We choose a test statistic that depends on the Mahalanobis distance between the mean point of subset and the mean point of the overall set. The p-value is the proportion of samples whose test statistic is more or as extreme as the observed one.

---

[7] Permutation tests were initiated by Fisher (1935) and Pitman (1937). For their use in GDA, see Le Roux et al. (2019).

A2.2 Geometric typicality test and compatibility zone

Here is a summary of the principle of the test. If the mean point of the group does not differ from the reference point O (null effect), each individual point of the group can be exchanged with the point symmetrical with respect to the reference point. For each permutation, the mean of values is calculated and the proportion of permutations for which the mean is more extreme than (or as extreme as) the observed mean defines the combinatorial p-value. The set of points compatible with the observed mean point at level alpha=0.05 define the 95% compatibility region which is delineated by an ellipse proportional to the concentration ellipse of the group.

A2.3 Homogeneity test

The principle of the test is the following. If two groups of individuals are homogeneous, we are free to exchange the observations between groups. In permutation theory, a set of *possible clouds* of the same structure is generated by exchanging points between the two sub-clouds (all arrangements of total points into two sub-clouds. With each possible cloud there is attached the deviation between the two mean points; the *test statistic* is the squared Mahalanobis distance between mean points, hence a permutation distribution of the test statistic (which is the analogous of a sampling distribution). The *combinatorial p-value* is the proportion of possible pairs of sub-clouds, for which the value of the test statistic is greater than (or equal to) that of the pair of sub-clouds under study.

A3: Tables

Table 3: The twelve active variables with the absolute frequencies of categories

1. Position in the field (2 indicators)	Position occupied in the GC	Executive board (EB) (n = 18)	Governor Non CB (n = 60)	Governor+EB (n = 7)
	Country group	German pole (n = 29)	Western pole (n = 26)	Southern pole (n = 30)
2. Education (3 indicators)	Educational level	Bachelor (n = 6)	Master (n = 30)	PhD (n = 49)
	Main discipline	Economics (n = 66)	Law (n = 11)	Management+Other (n = 8)
	Location studies abroad	no (n = 43)		
		Europe apart UK (n = 6)	USA (n = 21)	UK (n = 15)
3. Professional trajectory (7 indicators) with exclusive categories: made main career (main), made career (yes), did not make career (no)	Finance	main (n = 13)	yes (n = 26)	no (n = 46)
	Central bank (CB)	main (n = 26)	yes (n = 17)	no (n = 42)
	Private	main (n = 6)	yes (n = 19)	no (n = 60)
	Politics	main (n = 8)	yes (n = 17)	no (n = 60)
	University	main (n = 16)	yes (n = 24)	no (n = 45)
	Administration	main (n = 19)	yes (n = 28)	no (n = 38)
	IO	main (n = 3)	yes (n = 19)	no (n = 363)

Table 4: Interpretation of axis 1: 15 categories most contributing to axis 1

variables	Ctr (%)	left	right	Ctr (%)	categories
Country	11.7	western pole	southern pole	2.9	7.5
Educ.level	16.7	master	Phd	8.4	7.1
Educ.field	9.7	law/manag.		4.5+3.0	
Studies abroad	10.2	no	UK	4.4	4.1
CB	7.6	no	yes	3.3	3.7
University	14.4	no	main	5.3	8.6
Administration	15.9	main/yes	no	2.9+4.2	8.8

Sum of contributions: 78.6%

Table 5: Interpretation of axis 2: 13 categories most contributing to axis 1

variables	Ctr (%)	left	right	Ctr (%)	categories
Country	5.4	Western Pole		3.4	
Pos.GC	12.8	Gov+EB	EB	6.2	6.4
StudiesAbroad	8.6	Europe (apart UK)	USA	5.6	2.9
Finance	17.9	main	no	11.7	5.6
CB	13.7	yes	main	7.0	6.6
Private	7.3	main		4.9	
Politics	6.6		main		6.0
University	7.0	yes		4.9	
Administration	14.7	main		11.1	

Sum of contributions: 82.1%

Table 6: Interpretation of axis 3: 11 categories most contributing to axis 3

variables	Ctr (%)	left	right	Ctr (%)	categories
Country	5.2	German pole		3.3	
Pos.GC	6.6	Gov+EB		3.1	
Educ.level	7.1		Bachelor		2.8
Educ.field	21.4	Law	Manag+other	4.8	16.4
StudiesAbroad	11.7	Europe apart UK		9.5	
CB	5.2		main		3.1
Private	17.4		main		14.9
Administration	5.7	main		3.5	
IO	12.2	main/yes	no	9.1	3.2

Sum of contributions: 73.6%

Part II: Sociological applications of correspondence analysis and other multivariate scaling methods

Reconstructing a scientist's 'space of social relations' via multiple correspondence analysis

Alice Barth, Felix Leßke, Rebekka Atakan, Manuela Schmidt, Yvonne Scheit

Abstract

In this paper, we argue that social relationships and social networks are an important part of academia. We present a case study of a scientist's "space of social relations", based on survey data of 126 respondents who provided information about the duration and characteristics of their relationship to the German sociologist Jörg Blasius. Applying multiple correspondence analysis, we find that Jörg's space of social relations is structured by the contrast between a public/professional and a private/leisure pole, whereas the second dimension differentiates frequent from rare contact occasions. It is concluded that the analysis of social relations can provide insights into career paths and network structures in the social sciences

1. Introduction

Contrary to the popular metaphor of the 'ivory tower' and the Humboldtian ideal of scientific research "in solitude and freedom" (Humboldt 2010 [1809/10]: 229, own translation), social relationships are an integral part of any scientist's working life (Engler 2001). The role of collaborations is a factor that should not be underestimated for any successful scientific career (van der Waal et al 2021; Petersen 2015): "academic life [...] is embedded in social life" (McCabe 2020: 135).

Current research on social networks in science primarily focuses on the analysis of macroscopic collaboration patterns (Petersen 2015), i.e., the analysis of aggregate metrics of co-authorship, citations, or other formalized collaborations such as joint projects or membership in boards and commissions. A common source are bibliographic databases which are used to extract network data on citations and co-authorship (Newman 2004; Martin et al. 2013). Accordingly, the quality of relationships in scientific networks is quantified by indicators such as duration of collaboration and number of joint publications (Petersen 2015; Bu et al. 2018). Less is known, however, about the characteristics of scientists' networks beyond formal collaborations: How are relations characterized that do not result in joint projects or publications? What kinds of interactions take place apart from – or alongside – research activities? What are the structuring dimensions of individual social networks?

In this paper, we present the case study of one scientist's social network: Our example is Jörg, a German sociologist and statistician near retirement. In academia, a professor's retirement is, of course, not the end of their career as such, but means a change in responsibilities and activities; also, in this case at least, it provides the occasion for the current study, which was planned as a surprise to celebrate his lifetime achievements to date.

We collected survey data from more than 100 of Jörg's friends, colleagues, and collaborators, who answered questions on the nature of their relationship with him. By applying multiple correspondence analysis to this data, we aim to draw a picture of Jörg's current social network and analyzing its structure. Our approach is relational in a double sense: The variables used in the analysis describe the duration, quality, and nature of the respondents' relations to Jörg, while the analysis technique displays the relations of the variable categories and persons to each other in terms of patterns of similarity (or dissimilarity). The structure of the resulting low-dimensional 'space of social relations' for Jörg can be further interpreted by adding the social contacts' characteristics (age, gender, country of residence etc.) as supplementary variables.

2. On the importance of relations

Jörg is a person who values personal relations. For his research assistants, no conference passes without being introduced to his numerous friends and acquaintances, usually accompanied by comments such as "I first met her before you were even born" or "Whenever I meet him, we have a nice dinner and some glasses of wine". Both the duration of a relationship and the more pleasant aspects of scientific collaboration are of immediate relevance in his view. Consequently, the shift of academic collaboration from the physical to the digital space in the wake of the COVID-19 pandemic was all the more frustrating for Jörg; for him, meetings and conferences are not only a place of mutual scientific discussion, but above all a place where one can meet friends. Similarly, personal sympathy is an important attribute when Jörg talks about fellow scientists.

You can name almost anyone (who is within a certain age range, to be fair) in Jörg's various fields of research, and he will say "Oh, yes, I know him, he's a nice guy". Without ever "networking" on purpose, he has accumulated a huge circle of acquaintances in his scientific career, and many colleagues and collaborators have long become close personal friends.

In sociological theories, such social relations – both in the sense of personal friendships and of, for example, academic colleagues – are often conceptualized as "social capital"; however, common definitions of social capital differ quite widely (e.g., Adler & Kwon 2002). Robert Putnam defines it as mutual trust and shared norms that enable cooperation in society (Putnam 1994: 167). As such, social capital is a collective property that is seen as necessary for the functioning of democracies. Pierre Bourdieu, in contrast, highlights the material or symbolic resources that one can access by being a member of a social network (Bourdieu 2012[1983]: 238). In this sense, social capital is an individual asset – however, the full value of one's accumulated relations will be attained only when the purpose of establishing and cultivating a relationship is not oriented towards deriving a possible profit from it. The overt monetization of social capital directly undermines its actual efficacy.

Jörg's way of maintaining relationships can be considered an exemplary case of the accumulation of social capital without any ulterior motive. To him, it is a value in itself to be on friendly terms with (almost) everyone, irrespective of social position or the contact's potential power resources. Jörg's pronounced and non-strategic sociability makes him an ideal research subject for an analysis of the structure of his personal network, as a multitude of contacts happily agreed to answer survey questions about their relationship with him. The resulting data set might be referred to as a special case of an egocentric network: Whereas, usually, ego is asked to provide information about alteri and the nature of the relationships, in our study only the alteri are the source of information about their respective relationships.

3. Methodology

Rather than looking at the data with instruments of network analysis, for this study we – almost inevitably – decided to make use of multiple correspondence analysis. On the one hand, Jörg himself has been a leading scholar in the area of correspondence analysis and related methods for decades. On the other hand, we, the authors of this paper, are a group of Jörg's current and former PhD students and research assistants. Multiple years of working with him have accustomed us to a certain way of doing research (a scientific habitus, if you will). A core concept is a deep belief in exploratory data analysis following the dictum of the French statistician Jean-Paul Benzécri, one of the creators of cor-

71

respondence analysis, who stated "the model should follow the data, not vice versa" (Benzécri et al. 1973, own translation). Apart from these obvious reasons, multiple correspondence analysis is used for this study, as it is well-suited to displaying the similarities – and dissimilarities – in the relations that alteri have with Jörg. Therefore, it allows us to answer our overarching research question "What are the important structuring dimensions in Jörg's network of relations?"

A methodological preference often comes, be it conscious or not, with a certain theoretical orientation (Bourdieu 1979; Schmitz & Hamann 2022). In the case of multiple correspondence analysis, or techniques of geometric data analysis in general, many researchers have an affinity to the theoretical observations of Pierre Bourdieu. Jörg himself has extensively drawn on Bourdieusian concepts in his scientific career (e.g, Blasius & Winkler 1989; Blasius & Friedrichs 2003; Blasius & Mühlichen 2010; Blasius & Schmitz 2017). Thus, in this study, it was somewhat tempting to use Bourdieusian terminology in the analysis of Jörg's social network. While drafting the study, we called it 'Jörg's social space' – only to realize that, when thinking more carefully about it, the resulting low-dimensional representation of a single person's relationships does not correspond to the Bourdieusian conception of a 'social space'. Neither is it a representation of society, nor are "the active properties [...] selected as principles of construction of the social space [...] the different kinds of power or capital which are current in the different fields" (Bourdieu 1985: 196).

Undoubtedly, Jörg's social contacts have differential resources, both within the academic field (just compare PhD students on temporary contracts to established professors) and with regard to their cultural, economic, and symbolic capital in the social space, i.e. society in general. These resources will not, however, be analyzed as properties of the space in this study. In our case, the actors' relative positions are determined by the nature and quality of their relationship to Jörg – as such, the structuring dimensions of the resulting space should not be interpreted as "a set of objective power relations" (Bourdieu 1985: 196) or even (reader beware!) 'relationship capital'. Nevertheless, 'space' seems an appropriate term, as "[t]he notion of space contains, in itself, the principle of a relational understanding of the social world" (Bourdieu 1998: 31), which is why we decided to call the result of our analyses a 'space of social relations'.

4. Data and methods

In this study, we present the case of a specific scholar's social network. We focus on first-order contacts, that is, ego (Jörg) maintains direct contact with them or at least did so at a certain stage in his career. It is an ego-centered network, as we only assessed the nature of relationships between ego and alter, but not between the alteri. We will, however, assess shared characteristics of alteri via multiple correspondence analysis. While the usual approach in studies of egocentric networks is to question ego about the ties to alteri, due to the secret nature of the project, we had to contact alteri without the knowledge of ego. The ties we study are consensual in the sense that ego and alter are friends, or at least friendly acquaintances, and share certain values and interests (Gross 1956); while some ties do involve elements of institutional obligations (e.g., ego and alter are colleagues in the same department; ego was alter's PhD supervisor, etc.), contacts without a certain level of mutual sympathy were not included in the analysis.

The data for this study were collected via an online survey in February and March 2022. Approximately 160 of Jörg's colleagues, friends, and family members were personally contacted by the authors via e-mail; 126 respondents completed the survey. The study, despite the effort to integrate alteri from earlier career stages, reflects Jörg's current social network. Some former co-authors or colleagues could not be included because they have passed away or were unavailable for the survey, for instance if we could not locate a working e-mail address.

The main part of the survey treated the relationship of alteri to ego: the length of the relationship, where the first meeting took place, the frequency of contact between respondent and Jörg, the kind of relationship (e.g. (former) PhD student, conference buddy, friend), scientific topics he and the respondent have in common, and activities performed together (e.g., writing a paper together, having lunch, going to a party, etc.). In addition, we collected data on respondents' characteristics: their main scientific field (if applicable), gender, age group, and country of residence.

The data are analyzed using multiple correspondence analysis. Multiple correspondence analysis (MCA) is a scaling method for categorical data (Greenacre 2016). As such, it transforms a set of categorical observed variables into a smaller set of continuous latent variables (dimensions). For this purpose, singular value decomposition is used. Correspondence analysis has sometimes been referred to as principal component analysis with categorical data (Le Roux & Rouanet 2004: 180). Similar to principal component analysis, MCA provides eigenvalues ('contributions to inertia') and factor loadings. However, it is usually the graphic representation of categories (and individuals) in a low-dimensional Euclidean space that forms the main basis for interpretation. Spatial proximity in the geometric representation indicates similarity, i.e., categories that co-occur relatively often in the data are located close to each other in this map, whereas categories that do not or seldom co-occur are visualized far apart from each other. MCA enables the differentiation of active and passive (supplementary) variables. While active variables contribute to the geometric orientation of the axes, passive variables are projected into the space without changing its properties. Their spatial location, however, indicates how they are related to the active variables and can thus be an aid to interpretation.

In the first step of our analysis, we use relational variables as active variables in the construction of the social space. These are: place of the first meeting of Jörg and respondent; current frequency of contact; nature of their relationship (friend, colleague, etc.); scientific topics respondent and Jörg have in common; and activities performed together (even once, not necessarily frequently), such as having dinner, writing a paper together, going to a party. This approach will uncover the most important structural dimensions in the nature and quality of our respondents' relationships to Jörg. In a second step, the respondents' characteristics (age; gender; scientific field; country of residence) as well as the length of their relationship with Jörg are projected as supplementary variables into the space. This allows for a more detailed description of the properties of the different social groups identified in the space.

5. Results

Our sample consists of 126 persons. 81 identified themselves as male, 41 as female and two smaller groups of four people each said "other" or gave no answer, respectively. Table A1 in the appendix displays the age groups. Most of the respondents (73%) currently reside in Germany, 3% in the Netherlands, and 2% each in Austria, Switzerland, and France. A number of countries (Canada, Denmark, Hungary, Italy, Japan, Norway, Slovenia, South Africa, Spain, Turkey, the United Kingdom, and the United States), were each mentioned by one or two respondents. About half of the sample said that their main scientific discipline was sociology, 18% indicated statistics as their field, and 16% did not work as scientists. Other disciplines mentioned by small fractions of the sample were political science, economics, English studies, German philology, geography, and media studies.

Regarding the place of first meeting Jörg, 6% of respondents stated Hamburg (where Jörg grew up and studied), 28% Cologne (where he moved for his PhD), and 37% Bonn (where Jörg took up a professorship of sociology in 2001). 20% said that they met him at a conference or summer school in a different city. Respondents' frequency of meeting Jörg is reported in table A2 in the appendix.

Among the possible scientific topics that alteri shared with Jörg, methods of empirical social research was most often mentioned (62%), followed by statistics (47%), and correspondence analysis (37%). About a third of respondents shared research topics in the tradition of Pierre Bourdieu with Jörg; a similar number mentioned lifestyle research as a common scientific topic. Finally, quality of data was stated by 21% and gentrification by 17%.

The respondents were also asked about the nature of their relationship with Jörg – multiple responses were possible. Many considered themselves Jörg's friends (31%) or conference buddies (21%). There is also a substantial number of colleagues (26%) and collaborators (29%). Some respondents are (former) PhD candidates (17%), and/or (former) research assistants (18%). The questionnaire was also completed by six of Jörg's family members.

Regarding selected activities undertaken with Jörg, 89.6% of respondents answered that they had shared lunch or dinner, and 69.6% reported that they had visited a conference together. Scientific activities like giving a presentation (23.2%) or writing a paper together (16.8%) were less frequent, whereas 24% of respondents reported that they had stayed overnight at Jörg's place and 25.6% remembered going to a party together. Finally, respondents were asked about the beverages they consume when meeting him (here, red wine was the most prominent answer, being mentioned by more than half of the sample).

We relate these relational variables to each other by means of multiple correspondence analysis. We first show the space of social relations created from the active variables "place of first meeting", "scientific topics in common", "nature of relationship" (friend, colleague, PhD candidate and so on), "frequency of contact", and "activities performed together" (see figure 1). As the few family members that took part in the survey were identified as outlier cases, they were treated as "supplementary individuals" in the analysis who did not influence the orientation of the space.

In the upper right, there is a cluster of leisure activities that respondents engage(d) in with Jörg, like going on vacation, watching a movie, attending a concert, or playing a game. Also located in this quadrant are house concerts, parties, overnight visits, seeing oneself as a friend of Jörg, and Hamburg as the first meeting place. All in all, relationships in this area of the space are non-scientific and leisure-oriented. Jörg's family members, who were included as supplementary cases in the analysis, can be found far to the upper right as well (not shown).

Figure 1: Jörg's space of social relations, active variables

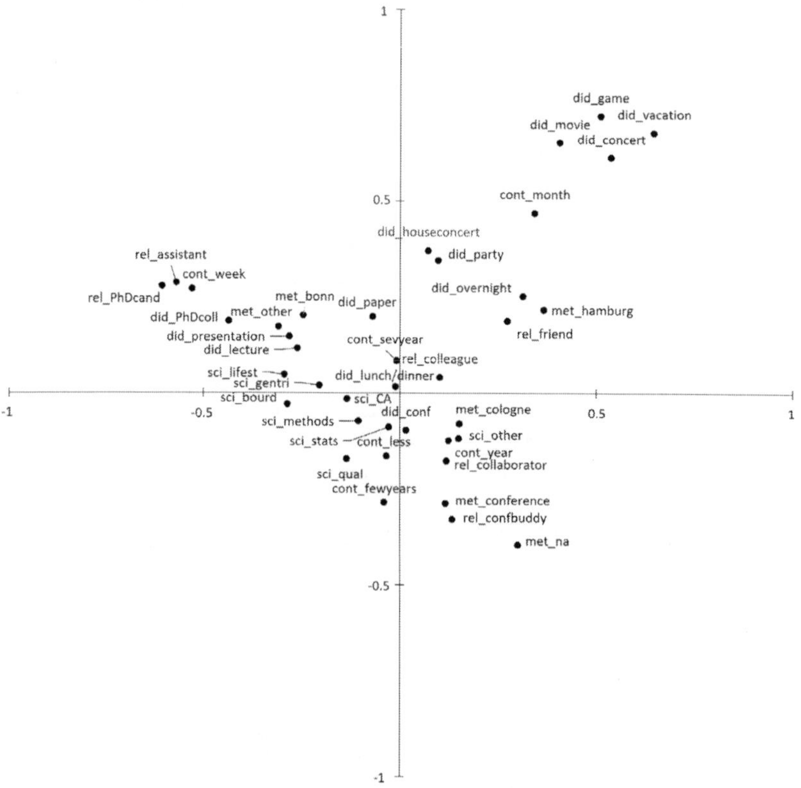

In the upper left quadrant, respondents are located who stated being Jörg's research assistants and meeting him once a week or more, as well as a number of study- and research-related activities such as attending a lecture or a PhD colloquium by Jörg, and giving a presentation together. The activity of writing a paper together is also situated at the upper left, albeit more to the middle. This kind of professional relationship usually started in Bonn and is associated with a number of shared scientific topics, such as lifestyles, Bourdieu, and gentrification.

Finally, in the lower part of the space, we can see more peripheral relationships, well described by the term "conference buddy". Here, respondents meet Jörg yearly, once every two to four years (the usual rhythm of conferences), or less often, and often the first meeting took place at a conference or summer school. Common interest in social science research methods and the quality of data, as well as the relationship status "collaborator", are also situated here. The activity "had lunch/dinner together" was mentioned by almost everyone and is accordingly located near the origin.

Thus, the structuring dimensions of Jörg's space of social relations can provisionally be interpreted as follows: The first – horizontal – dimension differentiates between more private, leisure-oriented relations on the right and professional, work-oriented relations on the left. The second dimension describes the intensity/frequency of contact: occasional meetings at the bottom, frequent collaboration or frequent private activities at the top.

Figure 2: Jörg's space of social relations, passive variables

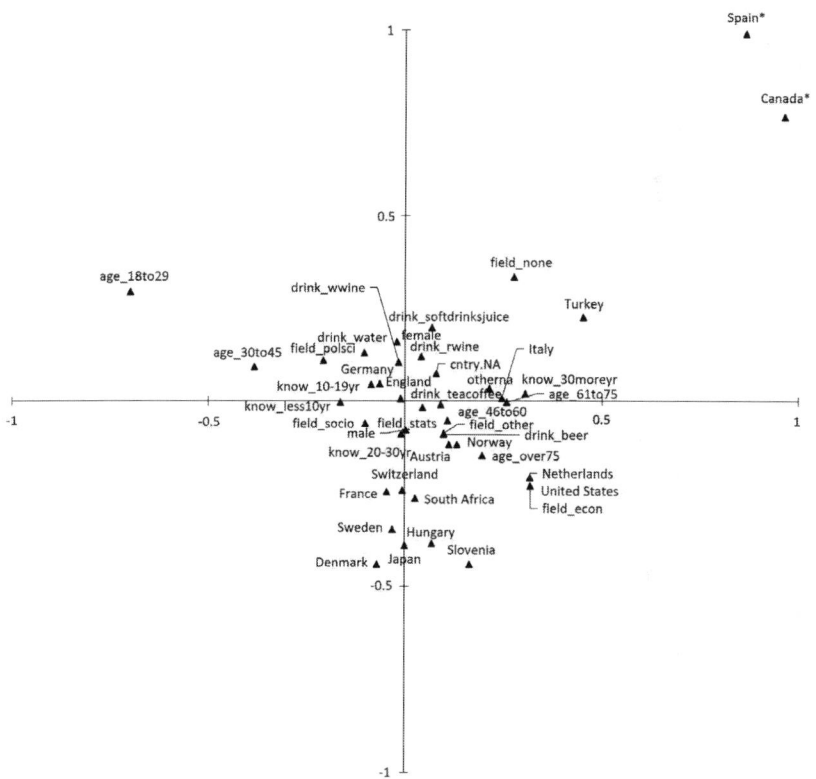

* indicates that category points were moved towards the centroid to improve readability

In the second step of the analysis, we enrich the interpretation of the space by adding passive variables that further describe the respondents and their relationship to Jörg (see figure 2). Respondents' age approximately runs along the first dimension, with younger contacts in the professional realm and older respondents in the friend/leisure area. The variable "length of relationship to Jörg" is similarly distributed, with relationships longer than 30 years on the right and less than 10 years on the left.

As could be expected, the statement "I am not a scientist" (scientific field: none) is located in the upper right as well, whereas sociology, statistics, economics and the respondents' other main scientific fields are in the lower part of the space, where scientific collaboration and meeting at conferences take place. The gender of contacts is differentiated on the second axis, with male contacts in the direction of occasional, conference-related meetings and female contacts at the top, suggesting a higher frequency of contacts. While most international contacts are located at the bottom, and thus represent "conference buddies", there are a few international contacts that are situated far to the upper right, and therefore the private leisure-side. The most frequent category, respondents who currently reside in Germany, is in the upper left quadrant, suggesting that the majority of German contacts were acquired in a later stage of Jörg's career. We also assessed what respondents usually drink when meeting Jörg: Beer is in the lower right quadrant and thus associated with less frequent meetings and older contacts, whereas water, red wine, white wine, and soft drinks/juice are in the positive part of the second axis and are therefore consumed in meetings that take place more frequently.

6. Discussion and Conclusion

In this paper, we aimed to assess the main structuring dimensions of a German sociologist's 'space of social relations'. We argued that personal relations are important in the scientific context in different forms. However, many analyses of scientific networks focus on purely professional collaborations, in terms of co-authorship, citations, and joint projects, while social relationships that are not directly 'used' for the generation of scientific output are often ignored by design. From a Bourdieusian perspective on social capital, however, it is particularly contacts that are maintained without a 'logic of exploitation' that ultimately strengthen one's position in a certain field.

Against this background, we collected data from 126 contacts, which mainly, but not exclusively, were acquired in Jörg's professional life. Our main variables of interest were respondents' descriptions of their relationship to Jörg, its nature, duration, starting point, and content in terms of activities engaged in together. The data were analyzed using multiple correspondence analysis, in order to create a 'space of social relations'. In this space, similar relationship attributes are in close proximity, whereas categories that were seldom mentioned together are distant from each other. In a second step, supplementary variables to describe respondents' characteristics were added.

The analysis showed that the main differentiation within Jörg's space of social relations was between a private sphere associated mainly with leisure activities, and a professional sphere with frequent, work-related interactions. Whereas respondents in the professional sphere are often (former) PhD candidates or research assistants, met Jörg in Bonn, and stated common interests in scientific topics such as gentrification and lifestyle research, contacts in the private sphere commonly use the word "friend" to describe their relationship and do not share specific scientific topics with Jörg. The second dimension separates frequent contact in private or professional relationships from respondents who see Jörg once a year or less often; the latter can mainly be described as "conference buddies" and scientific collaborators in the realm of methods of empirical social research and statistics. Respondents' characteristics also vary according to their position in Jörg's space of social relations; for instance, whereas younger respondents mainly maintain a professional relationship with a high frequency of contact, older respondents are predominantly located on the leisure side and/or are conference buddies.

In this sense, Jörg's space of social relations is indicative of a certain career path in academia, or, more specifically, the position of an established, nationally and internationally renowned professor: some acquaintances from an early career stage have become close friends that have a relational position similar to family members; Jörg has acquired a large circle of (international) collaborators and conference buddies, whereas day-to-day interactions at the office mainly occur with research assistants and PhD candidates. It is also interesting to see that Jörg's international collaborations are primarily in the fields of empirical social research and statistics, whereas studies concerning gentrification and lifestyles were conducted with German colleagues and assistants.

At the same time, the aforementioned relevance of sociability is a characteristic that can be found all across Jörg's space of social relations, not limited to the circle of family and friends. For example, 75% percent of the sample mentioned having had alcoholic drinks with Jörg – this number includes almost all of his research associates and (former) PhD students, partly owing to the practice of drinking wine at each PhD colloquium. This is a ritual Jörg adopted from his own PhD supervisor and one he would never forgo. He also values good food in pleasant company, which is reflected in the position of the category "had lunch/dinner together" in the center of the space, being mentioned by 90% of respondents. Here, Jörg's appreciation of the more pleasant aspects of academic life – and indeed life in general – is reflected in the data.

Among the study's limitations is the fact that the relevant contacts were not chosen by Jörg himself, but by the authors of the paper who, despite their best efforts, may have had some blind spots, particularly regarding social contacts from the earlier stages of Jörg's career. The same applies to the choice of variables in the survey. As current or former research assistants, the authors themselves are a part of Jörg's space of social relations; our subjective involvement in the field of research can, however, be resolved by the applied methodology: In the analysis of (social) fields or spaces, analysts will commonly, or almost invariably, occupy a certain position in the respective field or space, but the reconstruction of the relational structures by means of geometric data analysis can be seen as a means of objectivation (Blasius et al. 2019).

In summary, the relational analysis of social relations demonstrates the various facets of a social scientist's social circle, reflecting both the general structural aspects of contact formation in academia and the individual configuration of these contacts by a uniquely sociable and friendly person. Jörg, we wish you a good time in your well-deserved retirement!

References

Adler, Paul S./Kwon, Seok-Woo (2002): Social Capital: Prospects for a New Concept. In: Academy of Management Review 27, 1, pp. 17–40.

Benzécri, Jean-Paul, et al. (1973). L'analyse des données: L'analyse des correspondances. Paris: Dunod.

Blasius, Jörg/Friedrichs, Jürgen (2003): Les compétences pratiques font-elles partie du capital culturel? In: Revue française de sociologie 44, 3, pp. 549–576.

Blasius, Jörg/Lebaron, Frédéric/Le Roux, Brigitte/Schmitz, Andreas (2019): Investigations of Social Space: Introduction. In: Blasius, Jörg/Lebaron, Frédéric/Le Roux, Brigitte/Schmitz, Andreas (Eds.): Empirical Investigations of Social Space. Cham: Springer International Publishing, pp. 1–11.

Blasius, Jörg/Mühlichen, Andreas (2010): Identifying audience segments applying the "social space" approach. In: Poetics 38, 1, pp. 69–89.

Blasius, Jörg/Schmitz, Andreas (2017): Conference Report – Empirical Investigation of Social Space II University of Bonn, 12–14 October 2015. In: Bulletin of Sociological Methodology/Bulletin de Méthodologie Sociologique 133, pp. 65–70.

Blasius, Jörg/Winkler, Joachim (1989): Gibt es die „feinen Unterschiede"? Eine empirische Überprüfung der Bourdieuschen Theorie. In: Kölner Zeitschrift für Soziologie und Sozialpsychologie 41, 1, pp. 72–94.

Bourdieu, Pierre (1979): Public opinion does not exist. In: Mattelart, Armand/Siegelaub, Seth (Hrsg.): Communication and class struggle, 1. New York: International General, pp. 124–130.

Bourdieu, Pierre (2012 [1983]): Ökonomisches Kapital, kulturelles Kapital, soziales Kapital. In: Bauer, Ullrich/Bittlingmayer, Uwe H./Scherr, Albert (Eds.) Handbuch Bildungs- und Erziehungssoziologie, Wiesbaden: Springer VS, pp. 229–242.

Bourdieu, Pierre (1985): The social space and the genesis of groups. In: Social Science Information 24, 2, pp. 195–220.

Bourdieu, Pierre (1998): Practical reason. On the theory of action. Stanford, Kalifornien: Stanford University Press.

Bu, Yi/Murray, Dakota S./Ding, Ying/Huang, Yong/Zhao, Yiming (2018): Measuring the Stability of Scientific Collaboration. In: Scientometrics 114, pp. 463–479.

Engler, Steffani (2001): "In Einsamkeit und Freiheit?". Zur Konstruktion der wissenschaftlichen Persönlichkeit auf dem Weg zur Professur. Analyse und Forschung Sozialwissenschaften. Konstanz: UVK-Verlagsgesellschaft.

Greenacre, Michael J. (2016): Correspondence analysis in practice. Chapman & Hall/CRC interdisciplinary statistics series. Boca Raton: CRC Press, Taylor & Francis Group. Third edition.

Gross, Edward (1956): Symbiosis and consensus as integrative factors in small groups. American Sociological Review 21.2 pp. 174–179.

Humboldt, Wilhelm von (2010 [1809/10]): Über die innere und äussere Organisation der höheren wissenschaftlichen Anstalten in Berlin. Humboldt-Universität zu Berlin. Accessed at https://edoc.hu-berlin.de/bitstream/handle/18452/5305/229.pdf?sequence=1&isAllowed=y, December 5th 2022.

Le Roux, Brigitte/Rouanet, Henry (2004): Geometric data analysis. From correspondence analysis to structured data analysis. Dordrecht: Springer.

McCabe, Janice (2020): Why Study with Friends? A Relational Analysis of Students' Strategies to Integrate Social and Academic Life. In: Tierney, William G./Kolluri, Suneal (Eds.): Relational Sociology and Research on Schools, Colleges, and Universities. Albany: State University of New York Press, pp. 135–156.

Martin, Travis/Ball, Brian/Karrer, Brian/Newman, M. E. J. (2013): Coauthorship and citation patterns in the Physical Review. In: Physical review. E, Statistical, nonlinear, and soft matter physics 88, 1, p. 12814.

Newman, M. E. J. (2004): Coauthorship networks and patterns of scientific collaboration. In: Proceedings of the National Academy of Sciences of the United States of America 101, 1, pp. 5200–5205.

Petersen, Alexander M. (2015): Quantifying the impact of weak, strong, and super ties in scientific careers. In: Proceedings of the National Academy of Sciences of the United States of America 112, 34, pp. E4671-E4680.

Putnam, Robert D./Leonardi, Robert/Nanetti, Raffaella Y. (1994): Making Democracy Work. Princeton: Princeton University Press.

Schmitz, Andreas/Hamann, Julian (2022): The Nexus between Methods and Power in Sociological Research. In: The American Sociologist 53, 3, pp. 415–436.

van der Wal, Jessica E. M./Thorogood, Rose/Horrocks, Nicholas P. C. (2021): Collaboration enhances career progression in academic science, especially for female researchers. In: Proceedings of the Royal Society B. Biological sciences 288, 1958, 20210219.

Appendix

Table A1: Age groups of respondents

less than 18 years	4%
18–29 years	3%
30–45 years	24%
46–60 years	39%
61–75 years	20%
more than 75 years	9%

Table A2: Frequency of meeting Jörg

once a week or more often	8.7%
1–3 times a month	12.7%
several times a year	31.0%
once a year	18.3%
once in two to four years	16.7%
less often	12.7%

Religiosity and climate change beliefs in Europe
Martin Fritz & Yasemin El-Menouar

1. Introduction

In this chapter we investigate how religiosity and climate change beliefs are linked to each other. Why is this question relevant and worth studying? The answer is that religions are powerful social institutions which can and should play a crucial role in mobilizing society for the much-needed climate action. They can provide deeply rooted value foundations for a more ecologically sustainable mode of living and are able to reach out to great numbers of people. To tap this potential, it is important to understand the role of religion for human-nature relations and more specifically how different aspects of religiosity are connected to views on climate change.

Against the background of the increasing need and urgency of addressing climate change, most religions have taken official stances expressing their support for fast and far-reaching climate action. Probably most prominently in the Western world, the Laudato Si of Pope Francis (2015) addressed the topic, but there are similar statements or declarations in Islam and Buddhism, for example, as well as statements of interfaith organizations. While these pro-climate, pro-nature position-takings were declared at the level of institutions and led to a discussion about a potential 'greening' of religions (see 2.1), the question remains whether the adherents of religions share these views.

Previous empirical research which has investigated the links between religiosity and climate change beliefs, or more generally: views on nature, mainly focused on the US and other Anglo-Saxon countries, therefore analyzing primarily different Christian denominations. Applying methods of causal inference, these studies come to partly contradicting results: There are differences in the support for more climate and environmental protection between respondents of different religious denominations and between religious persons and non-believers. Some studies found Evangelicals to be least supportive; others show that Catholics are equally unconcerned about climate and ecological issues as Evangelicals. For the greening-of-religions hypothesis, only very little evidence is found in empirical research and there are indications that different aspects of religiosity such as attendance at religious services and praying have different effects on climate change beliefs. Also, interactions with other concepts like political ideology were shown. In short, the relationship between religiosity and climate change beliefs seems complex and largely remains unclear.

There is, however, a lack of quantitative research with a larger and different geographical focus than in existing studies. There is also a need for applying Correspondence Analyses and Related MEthods (CARME) to better explore the links between different aspects of religiosity and climate change beliefs without the methodological restrictions and assumptions that causal approaches have to deal with. This chapter presents a first attempt in closing this gap and proceeds as follows: We first discuss briefly how, on the one side, Judeo-Christian religions were criticized to promote human dominion over nature and an anthropocentric, dualist worldview and, on the other side, how researchers have argued about their recent greening (section 2.1). This is followed by a short overview on the existing studies on the relationship between religiosity and views on nature in general and climate change beliefs in particular (2.2). In section 3 we introduce our data, the European Social Survey, and our method, Multiple Correspondence Analysis, in more detail. The results, including graphical representations of the links between religiosity and climate change beliefs, are presented and discussed in section 4. Finally, we conclude the chapter with interpreting the results in light of religions' potential to mobilize for climate action and point to some possibilities for future research.

2. Theoretical reflections and previous empirical research on the links between religiosity and climate change beliefs

2.1 Changing religious stances towards nature: from domination to conservation?

Religion, particularly Judeo-Christian traditions, has been criticized by scientists for promoting human domination over nature. A key article making this argument was published by White (1967). Long before climate change entered the area of public attention he recognized the geological scope of ecological problems and states that fossil fuels threaten "...to change the chemistry of the globe's atmosphere as a whole, with consequences which we are only beginning to guess" (p. 1204). White argues that human relations to nature are "...deeply conditioned by beliefs about our nature and destiny – that is, by religion." (p. 1205). He attributes ecological destruction to the specific Judeo-Christian story of creation in which God created man in God's image and planned all the animals and plants for man's benefit and rule. Being an anthropocentric religion in contrast to ancient paganism and Asian religions, it would have established a dualism of humans and nature and insists that it is God's very will that humans exploit nature. Further, he demonstrates how this thinking has deeply shaped Western culture, economy and science and ultimately led to the 'great acceleration' of all kinds of human activities (Steffen et al. 2015) that today challenge us with multiple ecological crises.

Ecofeminist scholars also highlight that Western science, particularly in the tradition of Bacon, Hobbes and Descartes, facilitates controlling, manipulating and exploiting nature. These thinkers claimed the superiority of human reason over the chaos of nature and also referred to the Judeo-Christian story of creation: As humans were expelled from the garden of Eden and lost their rule over nature, it would now be the task of mankind to restore the lost dominion over nature by means of technology, science and reason (Merchant 1990, chapter 7). Moreover, Salleh (2017) shows that patriarchal religion replicates the dualism and hierarchical dominance between humans and nature in the social relationships between man and woman. Women are equated with nature and chaos while men are associated with reason and power. As both narratives amplify and support each other, authorities could easily justify men's rule over nature and women by means of control, manipulation and oppression.

Such critiques show how the Judeo-Christian religions in the West justified extensive interventions in nature including those causing climate change. But religious traditions globally are diverse and each well equipped with a variety of narratives that can be linked to very different and even opposing views of human nature relations. Therefore, it can be assumed that religious narratives are likely to correspond to those kinds of human nature relations that represent the spirit of the time: While on the eve of industrialization, scientific and technological nature domination brought progress and wealth for Western nations (at the cost of their colonies), this was justified and supported by religion. Today the negative ecological consequences often outweigh the benefits also in Western countries and citizens are concerned about the irrecoverable losses that are produced by ongoing nature exploitation. Are religions today therefore more critical of the dominion over nature and rather supportive of nature conservation? The so-called 'greening of religion' hypothesis says yes.

Koehrsen, Blanc, and Huber (2022) write "...White's contribution was an important starting point for academic debates about "ecology and religion" as well as for "greening" efforts within Christianity" (p. 46). Christian and other theologians started to develop "...ecological reinterpretations of their traditions and to generate faith-based environmental ethics that address the ecological crisis" (ibid.). In the literature these efforts have been interpreted as a 'greening' of religions, meaning that religions become more ecologically concerned over time (e.g. Chaplin 2016; Tucker 2008). Most of this 'greening' were official statements from religious leaders and networks like the Laudato Si from Pope Francis (2015) as well as the Islamic Declaration on Global Climate Change

(International Islamic Climate Change Symposium 2015) or the Interfaith Climate Change Statement to World Leaders which was signed by 176 religious groups from around the world (United Religions Initiative 2016). Simultaneously with the ecological awakening of religious institutions, researchers started to discuss the potential of religions to address environmental problems such as climate change. The main arguments for an ecologically progressive role of religions include their high outreach (over 80 percent of the world population belong to a religion), their ability to engage this broad audience through moral authority and leadership, and to influence the worldviews of their adherents with implications for the values and ethics that shape their lives. The economic resources and the political power that religions have at their disposal are mentioned as further advantages. Last but not least, their social capital (Putnam 1995) provides effective possibilities for mobilizing bottom-up environmental activities and achieving collective goals (Dilmaghani 2018; Haluza-DeLay 2014; Jenkins, Berry, & Kreider 2018; Koehrsen et al. 2022). Finally, the literature about religions' role for human-nature relations states that indigenous belief systems and new spiritualities foster greater concern about ecological problems and more support of environmental protection than the big world religions. It is argued that the reason is a holistic cosmology in which all beings are connected and the unity of all entities in nature including humans is emphasized (Koehrsen 2022; Taylor, Van Wieren, & Zaleha 2016).

Such theoretical reflections as well as the official statements and theologies show that there is a potential of religions to help mitigating and adapting to climate change. But what about the adherents and their beliefs themselves? Did they also become more concerned about ecological problems? Do their beliefs strengthen or suppress existing environmental consciousness? How are different aspects of their religiosity linked to climate change beliefs? Research investigating these questions is discussed in the next section.

2.2 Complex relationships between religiosity, nature and climate change beliefs

In a review of over 700 articles, Taylor et al. (2016) found that conservative theological orientations and beliefs about divine agency in natural events are main barriers for ecological concern and mobilization. They conclude that the thrust of White's thesis of the Judeo-Christian promotion of anthropocentric attitudes and nature destruction is supported, whereas the greening-of-religion hypothesis needs to be rejected.

Carlisle and Clark (2018) analyzed period and cohort effects on environmental views by religious affiliation with US data for the time 1973–2014. They found "...no clear evidence of a so-called "greening" of Christianity. Rather, our findings confirm studies that contend religion is not promoting pro-environmental behaviors" (p. 235). Contrary to their expectations, younger cohorts did not show more support for environmental spending. From their results, they conclude that religions contribute to ecologically destructive beliefs and practices. These findings were confirmed by Konisky (2018) who also analyzed US longitudinal data and found a decreasing concern about the environment among American Christians including Catholics, Protestants and other Christian denominations at differing levels of religiosity. Most recently, Koehrsen et al. (2022) summarize that: "...the environmental attitudes of religious followers do not reflect a potential 'greening' of their religions" (p. 48).

Moreover, there are studies on the climate change beliefs of Evangelicals who were assumed to be more skeptical about climate change and denying it more often than other believers and non-believers: Arbuckle and Konisky (2015) showed with data from the US that a negative association between being Christian and environmental attitudes was strongest among Evangelicals. In addition, they found an effect of religiosity: The more religious Evangelicals were, the lower was their concern about climate change. This was confirmed by a study of Veldman et al. (2021) who found that US Evangelicals were less likely to belief that climate change is real, that it is anthropogenic or that it is a serious problem than other religious groups. More specifically, they discovered that white Evangelicals are particularly skeptical of climate change, whereas Evangelicals of color were clearly less skeptical. These results may not only reflect a connection between notions of white supremacy and a religious justified dominion over nature, they also show that the link between Evangelical religiosity and climate change denialism may be affected by other factors than religion.

This assumption is supported by evidence from Brazil: Smith and Veldman (2020) showed that the environmental concern of persons with Evangelical and Pentecostal affiliation is not lower than for other denominations. The authors argue that "...laypeople often repurpose theological principles to support values they already hold" (p. 355). In the relatively pro-environmental public opinion context of Brazil this would lead Evangelicals and Pentecostals to justify their environmental values with reference to religious teachings. A greening of religion could therefore happen in the course of a value change towards greater concern for environmental protection. In this context, Lowe et al. (2022) analyze whether climate change attitudes of younger generations of US Evangelicals diverge from those of older generations. Their results show that younger Evangelicals more often than the older do not deny anthropogenic global warming. They even hold a range of pro-climate attitudes and support climate policies more than Americans overall. The authors suggest that a generational 'greening' of American Evangelicals may have taken place.

Other studies explore connections with political ideology to explain why sometimes religiosity is related to lower ecological concern and sometimes not (Haluza-DeLay 2014; Hayhoe, Bloom, & Webbä 2019). Jenkins et al. (2018) highlight that religious affiliation can soften the degree to which political ideology dictates climate opinion: Some aspects of religiosity such as frequency of religious attendance and strength of religious identity could ameliorate the negative effect of conservative political views on environmental behavior, while others such as biblical literalism intensify the negative effect (see p. 88). Peifer, Khalsa, and Ecklund (2016) find that religious attendance and identity are positively related to environmental consumption, whereas the belief in an involved God and biblical literalism are negatively related. Moreover, the authors assert religiosity would decrease the negative effect of political conservatism on environmental consumption. Arbuckle (2017) compares climate change concern of politically liberal and conservative persons between adherents of different religions and finds, in contrast to Jenkins et al. and Peifer et al., that the Evangelical, Black Protestant and Catholic traditions suppress climate concerns of liberals, whereas the Jewish tradition increases concerns. He argues: if "…a political position conflicts with religious eschatology, an otherwise politically liberal person might take the less liberal position, in this case, less concern about climate change" (p. 189). Dilmaghani (2018) used data from Canada and found that religiosity in general is a negative predictor of environmental money donations while religiosity, particularly religious attendance, is strongly positively related to environmental volunteering (donating time). The author explains these results by the causal contribution of religious attendance to the formation and maintenance of social networks (see p. 492).

Research on the relationship between religiosity and climate change beliefs with a special focus to other than Christian religions is sparse: In an article about Muslims perceptions of climate change Koehrsen (2021) emphasizes "…that the majority of Muslims regards climate change as an important societal challenge"(p. 5) but there is great variability between different countries and regions. He notes that quantitative research about how the religious worldviews of Muslims are linked to their views on climate change is still largely missing (see p. 7). The Jewish religion is often studied along with Christian denominations and mostly higher levels of concern about climate change are reported for Jews (Arbuckle & Konisky 2015). Orthodox religious groups such as Islamist Muslims (Yildirim, 2016) either engage in climate change denial or welcome it as divine punishment for human sins (Koehrsen et al. 2022, p. 48). Herman (2022) reports of ultra-orthodox Jews who practice avoidable bad ecological activities for religious reasons. A study with data from Australia (Morrison, Duncan, & Parton 2015) investigates differences between Buddhists, Christians and Secularists: Secularists were most concerned about climate change and more likely to believe in anthropogenic climate change, followed by Buddhists and both significantly more concerned than Christians. However, stronger explanatory factors were being female and higher educated. Controlling for these and other socio-demographic characteristics, Buddhists turned out more concerned than Secularists.

Most quantitative studies were conducted in Anglo-Saxon countries, particularly the US. For Europe we found only very few studies in a Web of Science search: An analysis of British survey data on environmental attitudes found no significant difference between Christians and non-Christians. Among Christians, the Roman Catholic were most skeptic toward nature. Interestingly, the authors also found that better education and more knowledge about nature increases attitudes of dominion over nature (Hayes & Marangudakis 2001). A recent study with data from Poland (Skalski et al. 2022) discovered that spirituality increases concern for the environment and the climate, whereas religious fundamentalism was negatively related to these variables. The relationship between religion and environmentalism was found mediated by right-wing authoritarianism which promotes anti-ecological views and practices. In addition, differences in the ecological views between Catholics and non-believers vanished when right-wing authoritarianism was included as mediator variable.

In summary, this short overview of existing research shows that the relationship between religiosity and climate change beliefs is complex and needs more research. In particular we found no studies which applied CARME and more research is needed with data outside the US covering different denominations including those beyond Christianity. The following analysis presents a first attempt to fill this gap.

3. Data and Methods

We used data from the European Social Survey (ESS) from the year 2016, edition 2.0. The survey covered 23 countries and contained questions on a variety of core topics repeated in all rounds of the survey, including questions on religiosity, and also a special module on public attitudes towards climate change. The survey applied strict random probability sampling and face-to-face interviews which were conducted between August 2016 and December 2017. The universe were persons aged 15+ years resident within private households in the participating countries (*European Social Survey edition 2.0* 2016). Table 1 provides an overview of the countries and sample sizes.

Table 1: Participating countries and sample sizes in the European Social Survey 2016

country	sample size	country	sample size
Austria	2 010	Lithuania	2 122
Belgium	1 766	Netherlands	1 681
Czech Republic	2 269	Norway	1 545
Estonia	2 019	Poland	1 694
Finland	1 925	Portugal	1 270
France	2 070	Russian Federation	2 430
Germany	2 852	Slovenia	1 307
Hungary	1 614	Spain	1 958
Iceland	880	Sweden	1 551
Ireland	2 757	Switzerland	1 525
Israel	2 557	United Kingdom	1 959
Italy	2 626	Total	44 387

To investigate people's religiosity, we used three questions from the survey: 1. Subjective religiosity "Regardless of whether you belong to a particular religion, how religious would you say you are?" measured on a 0=not at all religious to 10=very religious point scale. 2. Attendance of religious services "Apart from special occasions such as weddings and funerals, about how often do you attend religious services nowadays?" measured with the categories every day, more than once a week, once a week, at least once a month, only on special holidays, less often and never. 3. Frequency of praying "Apart from when you are at religious services, how often, if at all, do you pray?" with the same answering categories as the question on attendance.

We analyzed four questions to examine climate change beliefs: 1. Knowledge of the existence of climate change "You may have heard the idea that the world's climate is changing due to increases in temperature over the past 100 years. What is your personal opinion on this? Do you think the world's climate is changing?" measured with four categories definitely changing, probably changing, probably not changing, definitely not changing. 2. Knowledge of the causes of climate change "Do you think that climate change is caused by natural processes, human activity, or both?" with the answering options entirely by natural processes, mainly by natural processes, about equally by natural processes and human activity, mainly by human activity, entirely by human activity. 3. Personal responsibility to reduce climate change "To what extent do you feel a personal responsibility to try to reduce climate change?" measured on a 0=not at all to 10=a great deal point scale. 4. Personal worry about climate change "How worried are you about climate change?" measured on a 1=not at all worried to 5=extremely worried scale.

Some categories were merged to have a sufficient number of valid answers (at least five percent) for each category. We conducted a Multiple Correspondence Analysis (MCA). MCA is an explorative dimension reduction technique that analyses contingency tables of categorical data (Blasius 2001; Blasius et al. 2019; Greenacre 2007; Greenacre & Blasius 2006; Hjellbrekke 2019; Le Roux & Rouanet 2004).

> "It is often used to analyze survey data and aims to visualize similarities between respondents and relationships between categorical variables, and more precisely associations between categories." (Josse & Husson 2016, p. 15).

Most of our variables were categorical. The results of MCA are visualized in graphical representations that show how strongly and in which ways the variables are interrelated. We used the FactoMineR package (Lê, Josse, & Husson 2008) in R to conduct Multiple Correspondence Analysis. Moreover, the missMDA package (Josse & Husson 2016) was used to impute missing values by performing principal components methods on the incomplete data and under the missing at random assumption. The socio-economic characteristics age, gender, income and education as well as information on religious denomination and the country of the respondents were inserted as supplementary variables. In sum, the MCA included 37 categories within seven active variables. Further six supplementary variables were included.

4. Results

In this section, we first report the descriptive statistics for the seven questions used to assess respondents' religiosity and climate change beliefs. In the following, the statistical results of the MCA, the eigenvalues and most important contributions to the axes, are presented. Finally, we interpret the graphical representations generated by the MCA with regard to potential links between religiosity and climate change beliefs.

4.1 Description of the active variables

All degrees of subjective religiosity are present in the countries covered by the ESS 2016, albeit overall subjective religiosity seems rather moderate at most: About one third of the respondents reported high or very high religiosity (Table 2). Every fourth person evaluated her religiosity with some middle category (medium religiosity) and a relative majority of about 45 percent said they were not religious or only to a low degree.

Table 2: Subjective religiosity

How religious would you say you are?		
Category	Frequency	Percent
not at all (*religious* ---)	8 003	18.0
very low (*religious* --)	5 668	12.8
low (*religious* -)	5 979	13.5
medium (*religious* +)	10 729	24.2
high (*religious* ++)	9 394	21.2
very high (*religious* +++)	4 211	9.5
missing	403	0.9
Total	44 387	100.0

The social aspect of religiosity is reflected in one's participation in regular religious services. About one quarter of the respondents attended religious services once a month or more often (Table 3). Further 20 percent said they did so only on special days, another 20 percent seldom visited religious services and one third of the respondents reported they never participated in such events.

Table 3: Attendance of religious services

How often do you attend religious services nowadays?		
Category	Frequency	Percent
never (*rel. attendance* --)	15 708	35.4
seldom (*rel. attendance* -)	8 743	19.7
on special days (*rel. attendance* +-)	9 015	20.3
at least once a month (*rel. attendance* +)	4 373	9.9
weekly or more often (*rel. attendance* ++)	6 199	14.0
missing	349	0.8
Total	44 387	100.0

The most common practical expression of religiosity is praying, be it alone or together with other believers. Regarding their frequency of praying, the citizens of the countries included in the ESS 2016 were divided between a group of 18 percent who indicated daily praying and a group of more than twice as many persons who said they never pray (Table 4). Nearly six percent of the respondents reported to pray only on special holy days and 16 percent even less often.

Table 4: Frequency of praying

How often, if at all, do you pray?		
Category	Frequency	Percent
every day (*prayer +++*)	7 946	17.9
more than once a week (*prayer ++*)	3 273	7.4
once a week (*prayer +*)	2 636	5.9
at least once a month (*prayer +-*)	2 494	5.6
only on special holy days (*prayer -*)	2 494	5.6
less often (*prayer --*)	7 009	15.8
never (*prayer ---*)	17 559	39.6
missing	976	2.2
Total	44 387	100.0

The most important part of climate change beliefs is the belief or knowledge about whether the climate is actually changing. While science has shown that the climate is definitely changing (Intergovernmental Panel on Climate Change 2021), there were more than six percent among the respondents who indicated they believe the climate is definitely (2.2 percent) or probably (4.4 percent) not changing. These two categories have been merged in Table 5. While the opinion of these respondents could be close to those of politically active climate change deniers, the 36 percent who said the climate is probably changing seem more unsure about climate change. Slightly more than half of the respondents said that the climate is definitely changing, showing that the knowledge about the existence of climate change has diffused widely into the general population.

Table 5: Knowledge about the existence of climate change

Do you think the world's climate is changing?		
Category	Frequency	Percent
definitely changing	24 585	55.4
probably changing	15 764	35.5
not changing	2 940	6.6
missing	1 098	2.5
Total	44 387	100.0

However, the knowledge is much smaller when it comes to the question whether climate change is caused by humans or natural processes. There is overwhelming consensus among climate researchers that climate change is predominantly caused by anthropogenic CO_2 emissions that are released by burning fossil fuels

(Myers et al. 2021). 36 percent of the respondents were in line with scientific experts and believed climate change is mainly caused by human activity (Table 6). Seven percent of the respondents were even convinced that climate change is entirely caused by human activity. However, 43 percent believed natural processes and human activity contribute to climate change about equally and six to seven percent even said it is mainly natural processes which cause climate change.

Table 6: Knowledge on the causes of climate change

Do you think that climate change is caused by natural processes, human activity, or both?		
Category	Frequency	Percent
Mainly by natural processes	2 902	6.5
About equally by natural processes and human activity	19 126	43.1
Mainly by human activity	16 112	36.3
Entirely by human activity	2 954	6.7
missing	3 293	7.4
Total	44 387	100.0

Those respondents who did not totally deny the existence of climate change were asked whether they feel a personal responsibility to try to reduce climate change. In Table 7 the eleven categories of the answer scale were merged in such a way that new categories with about equal frequencies were created. In summary, there is a clear majority of the respondents who indicated to feel at least moderate personal responsibility – 55 percent chose 6 or higher on the 0–10-point scale.

The citizens of the mainly European countries covered by the ESS 2016 were mostly not very worried about climate change: Only one quarter of the respondents were very or extremely worried, 44 percent said they worry somewhat and another quarter was not very or not at all worried (Table 8). Today, probably more people would report to be worried as many catastrophic climate change related disasters such as forest fires, droughts, heat waves and floods have occurred in Europe since 2016.

Table 7: Personal responsibility to try to reduce climate change (categories merged from a 0 to 10 scale)

To what extent do you feel a personal responsibility to try to reduce climate change?		
Category	Frequency	Percent
no or very low responsibility (0–2) (*pers. responsibility cc ---*)	6 711	15.1
low responsibility (3–4) (*pers. responsibility cc --*)	5 241	11.8
low to medium responsibility (5) (*pers. responsibility cc -*)	6 739	15.2
medium responsibility (6) (*pers. responsibility cc +-*)	5 366	12.1
medium to high responsibility (7) (*pers. responsibility cc +*)	6 877	15.5
high responsibility (8) (*pers. responsibility cc ++*)	6 036	13.6
very high responsibility (9–10) (*pers. responsibility cc ++*)	4 957	11.2
missing	2 460	5.5
Total	44 387	100.0

Table 8: Worry about climate change

How worried are you about climate change?		
Category	Frequency	Percent
not at all (*cc worry --*)	2 590	5.8
not very (*cc worry -*)	8 493	19.1
somewhat (*cc worry +-*)	19 729	44.4
very (*cc worry +*)	9 649	21.7
extremely (*cc worry ++*)	2 193	4.9
missing	1 733	3.9
Total	44 387	100.0

4.2 Results of the MCA

Examination of the scree plot of the uncorrected eigenvalues suggested to consider the first three dimensions for further interpretation, as beyond this point the remaining eigenvalues are clearly smaller, of comparable size and declining constantly (Figure 1). A closer inspection of the category contributions to the third dimension revealed that this axis reflects the response behavior 'extreme categories vs. moderate categories'. It did not contain substantial information about the relationship between climate change beliefs and religiosity, therefore we focus on the first two dimensions in our discussion of the results.

As recommended in the literature, we calculated the Benzécri corrected eigenvalues for the first two dimensions to avoid underestimating the explained variance (Benzécri 1992; Hjellbrekke 2019; Le Roux & Rouanet 2004): The first dimension explains 49 percent of the total variance, the second another 25 percent.

Figure 1: Scree plot of the uncorrected eigenvalues

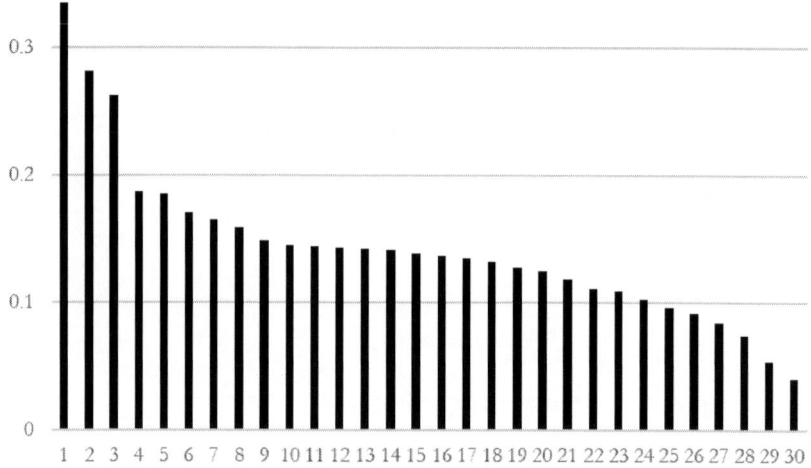

The first, horizontal axis reflects the degree of religiosity with low degrees on the left side and high degrees on the right. The categories with the highest contributions on the left side are: never praying (16.6) and never attending religious services (14.3) as well as being not religious at all (12.5). The highest contributions on the right side are 'praying every day' (11.4) and being highly and very highly religious (7.8 and 7.3).

The second, vertical axis represents a continuum of climate change beliefs where higher knowledge and greater worry are located at the bottom of the axis and climate skeptical and unworried views at the top of the axis. The highest contributions of variable categories here are: very high worry about climate change (10.8), very high felt responsibility (10.1) and the belief that the climate is definitely changing (9.3) at the bottom and no or very low felt responsibility (9.4) at the top of the axis.

4.3 Relationships between religiosity and climate change beliefs

The positions of all active variable categories relative to each other are depicted in the graphical representation of the first plane in Figure 2. It is striking that nearly all categories are very closely connected to 'their' axis. All categories operationalizing the different aspects of religiosity appear close to the horizontal axis and all climate change related categories stretch out over the vertical axis. This can be interpreted in two ways: First, the fact that the different aspects of religiosity and climate change, respectively, are highly associated to one axis indicates that there are underlying concepts of a general religiosity and climate change belief, in the sense of a latent trait: Persons with a high religiosity, for example, are likely to self-report a higher religiosity, to pray more frequently and to attend religious services more often than persons with a low religiosity. The three aspects are associated and form part of a general religiosity. The same can be said about the four aspects of climate change beliefs. Second, the fact that religiosity and climate change beliefs are nearly completely separated from each other and represented by different axes, suggests that the two concepts are independent from each other: Neither does a high religiosity necessarily include a disposition to be more climate skeptic or unworried about climate change nor does it automatically promote higher climate change awareness or feelings of responsibility to help tackling climate change. The only deviation from this otherwise very clear finding is the variable category 'praying only on special holidays' (prayer-). Although it contributes very little to each of both axes (less than 1%), it is also connected to the vertical axis and tends to the side with the skeptical views on climate change. A possible interpretation of this weak, yet outstanding association could be the following: Praying only on special holidays, which are central events of great importance in every religion celebrated by most believers, may reflect a conformist, social norms adhering behavior of those believers who are less religious in everyday life. Persons holding such a mind-set tend to avoid extremes and try to not offend other persons. Therefore, they would also give answers to climate questions that seem moderate to them given that they have less objective information. Indeed, the closest, most associated categories are 'the climate is probably changing' and 'climate change is equally caused by humans and nature'.

Figure 2: Religiosity and climate beliefs in Europe (MCA with Benzécri modified eigenvalues)

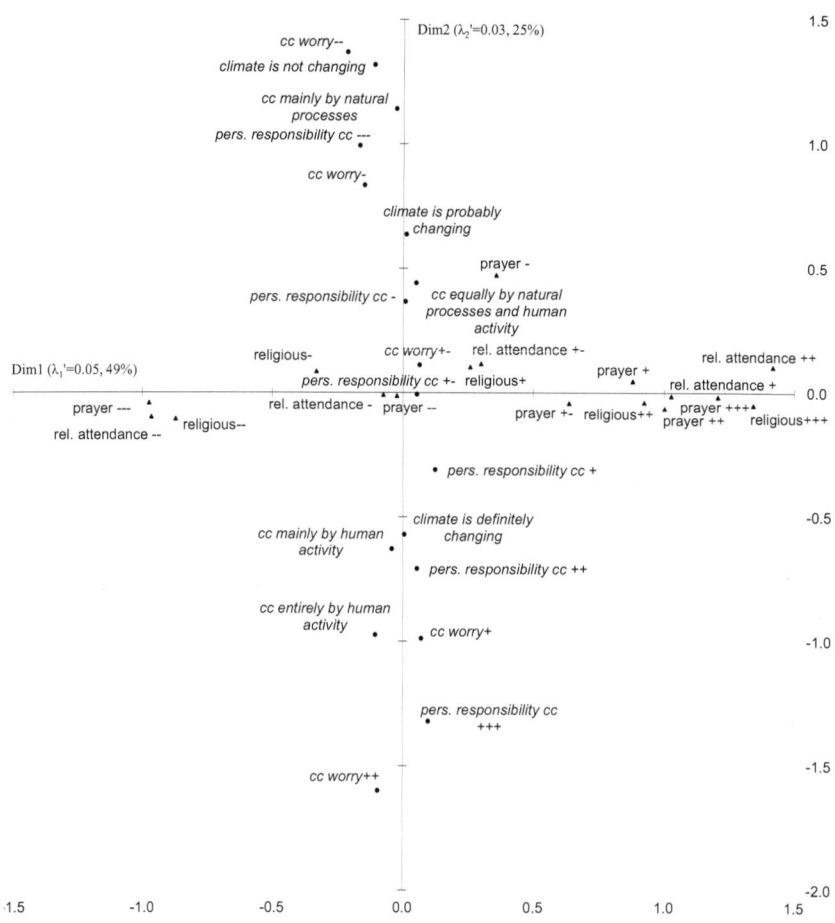

How are social characteristics and the different religious denominations linked to the two axes? In Figure 3 gender, age, income and education as well as denomination are inserted as supplementary categories. Men and women do not differ regarding their climate change beliefs, but women are more religious. Older persons are also more religious and more skeptical towards climate change than younger ones. Income is related to both axes: Persons with lower income tend to be more religious and skeptical towards climate changes, whereas persons with higher income are more likely to be aware of climate change and its anthropogenic causes and less religious. Education is stronger associated with climate change beliefs than with religiosity: A higher education is linked with climate change awareness, a lower education with some skepticism or unawareness and also with a higher religiosity. These findings resemble the results from many studies (see section 2 above) and validate our analysis.

Figure 3: Supplementary variables on the first plane of the MCA

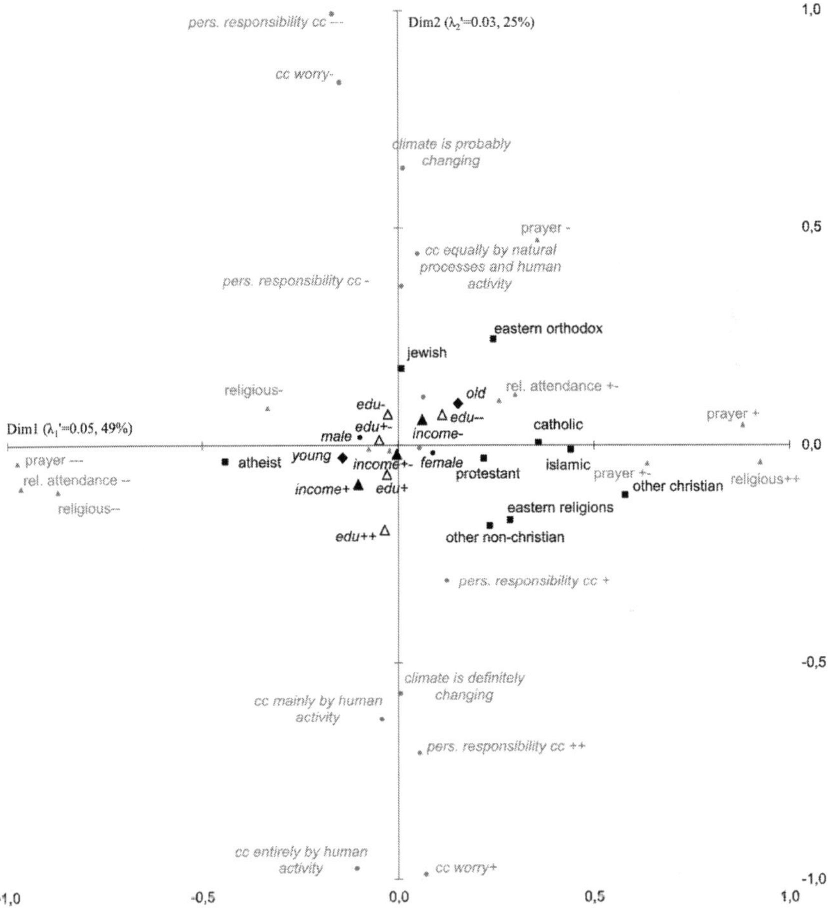

The positions of the different denominations included in the data are mostly in line with expectations and confirm previous studies in some aspects: First, it is logical that atheists are positioned most towards the left on horizontal axis, at the side of lower religiosity. The highest religiosity is found for persons who said they belong to another Christian denomination which includes, for example, the above discussed Evangelicals and Pentecostals. The second highest religiosity is found for Muslims, closely followed by Catholics. Eastern Orthodox and Protestant persons are about equally religious as well as Eastern religions and other non-Christian religions. Religious denominations differ with regard to their climate change beliefs: The most aware of climate change are Eastern and other non-Christian religions – a finding which confirms the above discussed study on Buddhists' climate change views. Other Christian religions (including Evangelicals) are also more aware and worried about climate change than the average. This adds further doubt to the argument that the rejection of ecological and climate concerns is higher among Evangelicals. Rather, it seems to have been an exceptional feature of US Evangelicals that recently started to dwindle (Lowe et al. 2022). Persons with an Eastern Orthodox and Jewish affiliation are on average more likely to be unconcerned and skeptical about climate change. While these findings contradict earlier studies for the case of Jews who were not less concerned than Christians or atheists (Arbuckle 2017; Arbuckle & Konisky 2015), in the case of Eastern Orthodox new empirical knowledge was created as they are rarely included in prior studies. It is unclear why both denominations are less concerned about climate change. It could be due to a Judeo-Christian anthropocentric and dualist worldview as White (1967) originally argued, but then the other Christian denominations do not fit in the explanation. More likely seems that other factors like the geographical and thus cultural and political distribution of the denominations play a role: Most Jews in the sample of the ESS live in Israel and most Orthodox persons in post-socialist Eastern European countries. Climate change awareness and support for ecological concerns are known to be lower in these countries (Fritz & Koch 2019). Figure 4 confirms this and shows the positions of all countries in the sample. There are countries in each of the four quadrants: Countries with a comparatively low religiosity and higher climate change skepticism are Russia, Estonia and the Czech Republic, all are countries with considerable shares of atheists and a high importance of the fossil industry for their national economy which promotes climate change denial and rejection of climate policies. Rather religious and skeptical towards climate change are the citizens in Poland, Lithuania and Israel. Among the countries with the highest average climate change awareness and concern, Portugal is the most religious while France as well as Germany feature a relatively low religiosity. Overall, the cloud-like distribution of countries in the first plane illustrates that all combinations of religiosity and climate change beliefs do occur and there is no sign of a connection between religiosity and climate change on the country level.

Figure 4: Positions of countries on the first plane of the MCA

5. Conclusions

This chapter analyzed the relationship between religiosity and climate change beliefs by means of a Multiple Correspondence Analysis. We discussed how religions can be theorized to support dominion over nature on the one side and care for nature as God's creation on the other side, and asked whether and how these rather theological questions echo in individual religiosity. Reviewing previous studies on the links between religiosity and climate change beliefs, a complex picture emerged in which denominations, different aspects of religiosity and social characteristics interact to affect peoples' views on climate change. We noticed a lack of research with large survey data outside the US and an absence of explorative methods that discover and visualize the relationships between different aspects of religiosity and climate change beliefs without the methodological restrictions and assumptions of causal approaches. In our analysis we used data from the ESS covering 23 countries and a wide range of different denominations such as Protestants and Catholics, Eastern Orthodox and Jews as well as Eastern religions and Islam. We applied MCA to 37 active categories representing respondents' religiosity and climate change beliefs. The key finding is that there is no straight-forward relationship between the two concepts. Both appeared as independent from each other.

Our results have also shown that denominations differ in their climate change views. Compared with previous studies, our findings highlight the role of context: While the more pro-climate Jews in the US are politically rather liberal and in socio-economic relatively privileged positions, Jews living in the unstable political situation of Israel turned out to be less concerned about climate change than other denominations. Future studies could use CARME to take a deeper look into the links between religiosity, political factors and climate change beliefs, or more generally human-nature-relations. Moreover, different understandings of religiosity within each denomination should be explored to compare orthodox, traditional and progressive interpretations of faith (El-Menouar 2014). Similar to prior research, we found that Eastern and non-Christian religions tend to be the most concerned with climate change. The notion that their holistic and non-dualistic worldviews make them more environmentally conscious may hold merit, but this hypothesis requires further investigation using data from regions where these religions are more widespread.

In terms of religion as a potential driver for enhancing climate action, our findings suggest the following: 1. Religiosity is not a hindrance to pro-climate views. 2. At the same time, it is not inherently linked to a pro-climate perspective. 3. The relationship between religiosity and views on climate change is influenced by a variety of factors which require further research. In the meantime, the lack of a clear relationship between religiosity and climate change beliefs opens up an opportunity for religious leaders to leverage their influence to stress the importance of preserving a habitable planet for current and future generations.

References

Arbuckle, M. B. (2017): The Interaction of Religion, Political Ideology, and Concern About Climate Change in the United States. In: Society & Natural Resources 30, 2, pp. 177-194.

Arbuckle, M. B./Konisky, D. M. (2015): The Role of Religion in Environmental Attitudes. In: Social Science Quarterly 96, 5, pp. 1244-1263.

Benzécri, J.-P. (1992): Correspondence Analysis Handbook. New York: Dekker.

Blasius, J. (2001): Korrespondenzanalyse. München: R. Oldenbourg Verlag.

Blasius, J./Lebaron, F./Le Roux, B./Schmitz, A. (Eds.). (2019): Empirical Investigations of Social Space. Cham: Springer.

Carlisle, J. E./Clark, A. K. (2018): Green for God: Religion and Environmentalism by Cohort and Time. In: Environment and Behavior 50, 2, pp. 213-241.

Chaplin, J. (2016): The global greening of religion. In: Palgrave Communications 2, 1, pp. 1-5.

Dilmaghani, M. (2018): Which is greener: secularity or religiosity? Environmental philanthropy along religiosity spectrum. In: Environmental Economics and Policy Studies 20, 2, pp. 477-502.

El-Menouar, Y. (2014): The Five Dimensions of Muslim Religiosity. Results of an Empirical Study. In: methods, data, analyses 8, 1, pp. 53-78.

European Social Survey Round 8 Data (2016). Data file edition 2.0. Sikt – Norwegian Agency for Shared Services in Education and Research, Norway – Data Archive and distributor of ESS data for ESS ERIC. doi:10.21338/NSD-ESS8-2016.

Fritz, M./Koch, M. (2019): Public Support for Sustainable Welfare Compared: Links between Attitudes towards Climate and Welfare Policies. In: Sustainability 11, 15, pp. 1-15.

Greenacre, M. (2007): Correspondence Analysis in Practice. Boca Raton: Chapman & Hall.

Greenacre, M./Blasius, J. (Eds.). (2006): Multiple Correspondence Analysis and Related Methods. Boca Raton: Chapman & Hall.

Haluza-DeLay, R. (2014): Religion and climate change: varieties in viewpoints and practices. In: WIREs Climate Change 5, 2, pp. 261-279.

Hayes, B. G./Marangudakis, M. (2001): Religion and attitudes towards nature in Britain. In: The British Journal of Sociology 52, 1, pp. 139-155.

Hayhoe, D./Bloom, M. A./Webb, B. S. (2019): Changing evangelical minds on climate change. In: Environmental Research Letters 14, 2, pp. 1-15.

Herman, L. (2022): Kosher Electricity and Sustainability. In: J. Koehrsen, J. Blanc, & F. Huber (Eds.), Religious Environmental Activism: Emerging Conflicts and Tensions in Earth Stewardship: Routledge, chapter 11.

Hjellbrekke, J. (2019): Multiple Correspondence Analysis For The Social Sciences. Abingdon, Oxon; New York, NY: Routledge, Taylor & Francis Group.

Intergovernmental Panel on Climate Change. (2021): Summary for Policymakers. In: V. Masson-Delmotte, P. Zhai, A. Pirani, S. Connors, C. Péan, S. Berger, N. Caud, Y. Chen, L. Goldfarb, M. Gomis, M. Huang, K. Leitzell, E. Lonnoy, J. Matthews, T. Maycock, T. Waterfield, O. Yelekçi, R. Yu, & B. Zho (Eds.), Climate Change 2021: The Physical Science Basis. Contribution of Working Group I to the Sixth Assessment Report of the

Intergovernmental Panel on Climate Change. Cambridge, United Kingdom and New York, NY, USA: Cambridge University Press, pp. 3–32.

International Islamic Climate Change Symposium. (2015): Islamic declaration on global climate change.

Jenkins, W./Berry, E./Kreider, L. B. (2018): Religion and Climate Change. In: Annual Review of Environment and Resources 43, 1, pp. 85–108.

Josse, J./Husson, F. (2016): missMDA: A Package for Handling Missing Values in Multivariate Data Analysis. In: Journal of Statistical Software, 70, pp. 1–31.

Koehrsen, J. (2021): Muslims and climate change: How Islam, Muslim organizations, and religious leaders influence climate change perceptions and mitigation activities. In: WIREs Climate Change 12, 3, pp. 1–19.

Koehrsen, J. (2022): Religion and Ecology. In: L. Pellizzoni, E. Leonardi, & V. Asara (Eds.), Handbook of Critical Environmental Politics. Edward Elgar, pp. 282–294.

Koehrsen, J./Blanc, J./Huber, F. (2022): How "green" can religions be? Tensions about religious environmentalism. In: Zeitschrift für Religion, Gesellschaft und Politik 6, 1, pp. 43–64.

Konisky, D. M. (2018): The greening of Christianity? A study of environmental attitudes over time. In: Environmental Politics 27, 2, pp. 267–291.

Le Roux, B./Rouanet, H. (2004): Geometric Data Analysis. Amsterdam: North Holland.

Lê, S./ Josse, J./Husson, F. (2008): FactoMineR: An R Package for Multivariate Analysis. In: Journal of Statistical Software 25, 1, pp. 1–18.

Lowe, B. S./Jacobson, S. K./Israel, G. D./Kotcher, J. E./Rosenthal, S. A./Maibach, E. W./Leiserowitz, A. (2022): The generational divide over climate change among American evangelicals. In: Environmental Research Letters 17, 11, pp. 1–13.

Merchant, C. (1990): The Death of Nature: Women, Ecology, and the Scientific Revolution. New York: HarperOne.

Morrison, M./Duncan, R./Parton, K. (2015): Religion Does Matter for Climate Change Attitudes and Behavior. In: Plos One 10, 8, pp. 1–16.

Myers, K. F./Doran, P. T./Cook, J., Kotcher, J. E./Myers, T. A. (2021): Consensus revisited: quantifying scientific agreement on climate change and climate expertise among Earth scientists 10 years later. In: Environmental Research Letters 16, 10, pp. 1–10.

Peifer, J. L., Khalsa, S./Ecklund, E. H. (2016): Political conservatism, religion, and environmental consumption in the United States. In: Environmental Politics 25, 4, p. 661–689.

Pope Francis. (2015): Laudato Si. On care for our common home.

Putnam, R. D. (1995): Bowling alone: America's declining social capital. In: Journal of Democracy 6, 1, pp. 65–78.

Salleh, A. (2017): Ecofeminism as Politics: Nature, Marx and the Postmodern (2nd ed.). London: Zed Books.

Skalski, S. B./Loichen, T./Toussaint, L. L./Uram, P./Kwiatkowska, A./Surzykiewicz, J. (2022): Relationships between Spirituality, Religious Fundamentalism and Environmentalism: The Mediating Role of Right-Wing Authoritarianism. In: International Journal of Environmental Research and Public Health 19, 20, pp. 1–11.

Smith, A. E./Veldman, R. G. (2020): Evangelical Environmentalists? Evidence from Brazil. In: Journal for the Scientific Study of Religion 59, 2, pp. 341–359.

Steffen, W./Broadgate, W./Deutsch, L./Gaffney, O./Ludwig, C. (2015): The trajectory of the Anthropocene: The Great Acceleration. In: The Anthropocene Review 2, 1, pp. 81–98.
Taylor, B./Van Wieren, G./Zaleha, B. D. (2016): Lynn White Jr. and the greening-of-religion hypothesis. In: Conservation Biology 30, 5, pp. 1000–1009.
Tucker, M. E. (2008): World Religions, the Earth Charter, and Sustainability. In: Worldviews: Global Religions, Culture, and Ecology 12, 2/3, pp. 115–128.
United Religions Initiative. (2016): Interfaith Climate Change Statement.
Veldman, R. G./Wald, D. M./Mills, S. B./Peterson, D. A. M. (2021): Who are American evangelical Protestants and why do they matter for US climate policy? In: WIREs Climate Change 12, 2, pp. 1–21.
White, L. (1967): The Historical Roots of Our Ecologic Crisis. In: Science 155, 3767, pp. 1203–1207.
Yildirim, A. K. (2016): Between anti-westernism and development: political Islam and environmentalism. In: Middle Eastern Studies 52, 2, pp. 215–232.

Health inequality, working conditions in dual vocational training and educational inequality – An analysis using categorical principal components analysis and hierarchical cluster analysis

Rahim Hajji, Simone Pollak, Gunnar Voß, Ulrike Scorna, Jessica Schäfer

1. Introduction

In 2021, there were a total of 1,255,440 trainees in Germany enrolled in approximately 330 recognised training occupations (Statistisches Bundesamt 2022; Bundesinstitut für Berufsbildung 2021). Vocational training in Germany provides young people with an institutionalised transition from school to the labour market. The more practical and application-orientated in-company vocational training does not require a higher school-leaving qualification. Vocational training can take two forms: dual training or school training. In dual vocational training, trainees are trained both in the company and at vocational school. As a general rule, dual vocational training takes 3 to 4 years to complete (Kaminski, Nauerth & Pfefferle 2008).

Countries with a work-centred vocational education and training (VET) system, such as Germany, have a comparatively lower unemployment rate among adolescents than countries without a work-centred VET system (Gessler 2017; Maaz et al. 2014). Furthermore, by teaching vocational skills, the work-centred VET system contributes to the social integration of adolescents in the workforce. The VET system also reduces the poverty risk in young people and promotes the economic success of companies (Busemeyer & Trampusch 2012; Li & Pilz 2021).

The integration of adolescents into the labour market through vocational training has the advantage of rapid occupational integration, but also the disadvantage that young people are exposed to occupational health risks (Duc & Lamamra 2022). Health risks could include negative physical effects from too much heavy lifting, psychological strain from burnout and excessive demand, and negative social risks such as experiencing a lack of integration in the workforce and appreciation by colleagues/ supervisors. These factors are all dependent on the working conditions in the training.

It is known from research that access to and placement in training depends on the school-leaving qualification and social background of adolescents (Protsch & Solga 2016). The social background of young people includes the educational background of their parents. According to this, adolescents with a comparatively higher school-leaving qualification and a better grade-point average who come from a household with a higher educational background are more likely to attain a more attractive and prestigious training place, characterised by a higher quality of training and a lower contract termination rate (Granato & Ulrich 2014). Duc and Lamamra (2022) point out in their study that often trainees terminate the training contract due to health burdens experienced at the workplace. In particular, trainees with a low level of education terminate their training contracts more frequently (Rohrbach-Schmidt & Uhly 2015). The contract termination rate is an estimate of the proportion of training contracts that are terminated. This rate increased from 23.0% in 2010 to 25.1% in 2022 (Bundesinstitut für Berufsbildung 2022).

It is obvious to ask to what extent health inequalities, working conditions in training and educational inequalities are interrelated among adolescents. There are very few studies on health inequalities and working conditions in training (Duc & Lamamra 2022), but there are numerous studies examining training opportunities/paths and educational inequality (Granato & Ulrich 2014; Maaz, Neumann & Baumert 2014; Protsch & Solga 2016). Consequently, there is a lack of studies that analyse the connection between health inequality, working conditions in training and educational inequality.

This study is structured as follows: Firstly, the aim of the present study is to bring together health inequality with working conditions in training and educational inequality in a theoretical framework in order to formulate empirically examinable correlation hypotheses. Secondly, the methodological approach for the study is presented. This is followed by the presentation and discussion of the results of a categorical principal components analysis to explore the relationships between health inequalities, working conditions and educational inequalities. The use of cluster analysis facilitates the identification of risk groups. Finally, empirical conclusions are summarised, critically reflected upon and a prospect on further studies is given.

2. State of the art and hypotheses

Health inequality is viewed as a consequence of social inequality (Siegrist & Dragano 2020; Mielck 2012). According to this, different working, housing and living conditions are characteristics of social inequality, which influence health. By studying social inequality and health, the extent of health inequality caused by social conditions can be measured and thus explained.

The World Health Organization (WHO 2010) has developed a framework model to illustrate and explain health equity, which we used to develop a model to explain health inequalities (see figure 1). We assume that the socio-economic and political context of a country affects health through governance and societal values. In this context, the welfare state plays a crucial role in the distribution of health opportunities. This is because the welfare state influences the living conditions and thus the health conditions of vulnerable population groups through the redistribution of financial resources in favour of people with reduced financial means.

Figure 1: Adapted WHO Framework Model (WHO 2010)

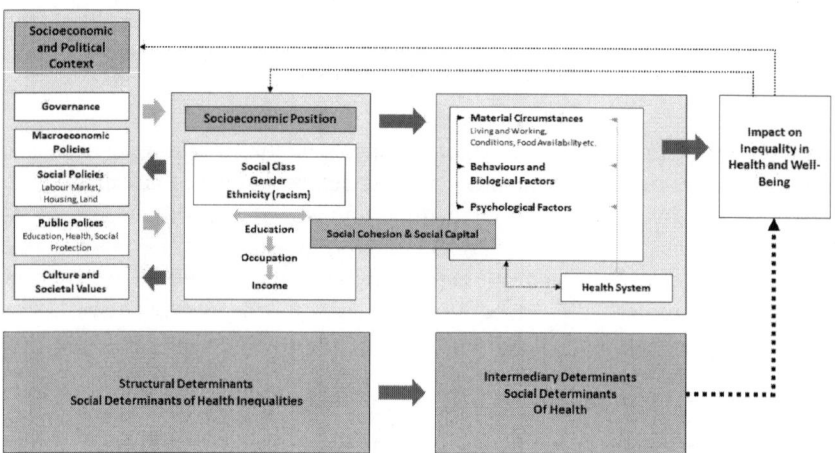

According to the model, the structural socio-economic and political conditions contribute to the social stratification of society. The socio-economic position of individuals can be described using income, education, occupational status, gender, ethnicity and social classes. The model assumes that due to educational qualification an occupation and thus an income is secured. Living and working conditions are related to socio-economic position. The model assumes that living and working conditions are associated with different work and environmental burdens, resources for housing and consumption, psycho-social conditions such as stressors and behavioural patterns. These factors then influence the distribution of health opportunities and thus lead to health inequalities.

The Job Demands-Resources Model (Bakker & Demerouti 2007) offers a holistic approach to explain health inequalities at the workplace. It is a theoretical model that can be applied to different settings. In a simplified way, the model assumes that working conditions at the workplace can be described in terms of job demands and resources, and that these terms can be used to explain health consequences. Job demands are understood to be all forms of physical, psychological and social stressors associated with physical or mental effort. Due to its flexibility, the model can account for very different job demands. Schaufeli and Taris (2014) characterise job demands as excessive demands, unfavourable working conditions, time pressure and job insecurity. They define job resources as organisational, social and psychological factors that are useful in achieving work goals, reducing job demands and promoting personal development. Within the term resources, the two authors include job control as well as social support from superiors and colleagues.

According to Bakker and Demerouti (2007), workload resulting from working conditions can be explained by demands and resources (see table 1). With high job demands and a high level of resources, the two authors expect a low to average workload. They justify this with the fact that the resource of job control gives employees possibilities to control the job demands. As a further example, the authors cite the presence of social support from colleagues and superiors as a factor which can reduce job demands and promote motivation. When job demands are low and resources are high, they conclude that work stressors are low. Therefore, work engagement is high due to the experience of social support and recognition from colleagues and supervisors. In settings where there are low job demands and low resources, they conclude that work stressors are low and work engagement is average. In settings with high job demands and low resources, workload is high and work engagement is low (Bakker & Demerouti 2007).

Table 1: Connection between job demands, resources, workload and motivation

	Low job demands	High job demands
Low resources	Low workload // average motivation	High workload // poor motivation
High resources	Low workload // strong motivation	Low to average workload // strong motivation

According to Schlicht (2011), educational inequality refers to the differences in educational behaviour and in the educational qualifications achieved, which can be explained by social inequalities in families and living conditions. The educational success of children and young people is often related to the socio-economic status of their parents. According to Hillmert (2011), social inequality in families of origin regarding occupation, income and education first translates into social inequality in education systems. Children with parents who have a high level of education are more likely to acquire the Abitur (the highest school-leaving qualification in Germany) than children with parents who have a low level of education. According to Becker and Lauterbach (2016), there are various reasons for this. Children and adolescents growing up in families with a higher socioeconomic status develop linguistic and cognitive skills that are advantageous for school, through socialisation and individual support in the family household. In addition, parents with a high socioeconomic status, generally with a higher level of education, increase the likelihood that children and adolescents strive towards higher educational paths and qualifications (Becker & Lauterbach 2016).

According to Hillmert (2011), social inequality in educational success relates to educational and labour market opportunities, resulting in social inequality in employment. He points out that children's access to and level of education depends on their schooling, which in turn depends on the educational level of their parents (Hillmert 2010, 2011). Granato and Ulrich (2014) state that companies are guided by school performance in the pre-selection of potential trainees. This is because, according to Granato and Ulrich (2014), the companies derive the potential trainees' performance from their school performance.

In summary, the health, labour and educational approaches can be graphically represented in an explanatory model (see figure 2).

Figure 2: Explanatory model on educational inequality, working conditions in education and health inequality

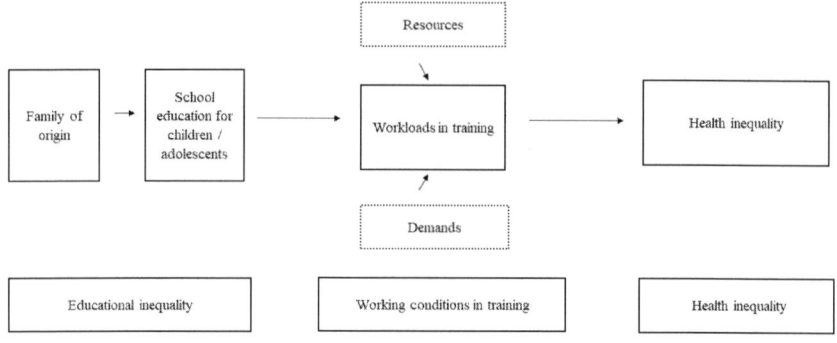

The present study aims to investigate whether educational inequality, working conditions in training and health inequality are interrelated. The study picks up Hillmert's (2014) thesis, who assumes that "the process of social reproduction", starting from the family of origin, leads firstly to social inequalities in the education system and then secondly to social inequalities in the employment system. In this case, Hillmert also speaks of the "inheritance of social inequalities".

We assume that educational inequality, working conditions in training and health inequality are interrelated. Adolescents with parents who have a high level of education are more likely to perform better at school than adolescents with parents who have a low level of education. As a consequence, young people with higher school-leaving qualifications and good school grades are more likely to choose better training occupations and companies. Consequently, they tend to experience better working conditions throughout their training and lower workloads at their workplace. This is in contrast to young people with low school-leaving qualifications and poorer school grades. Accordingly, adolescents with a high school-leaving qualification are more likely to complete their training in companies with a high level of resources and with low to high job demands, while adolescents with a low school-leaving qualification are likely to find themselves in training companies with a low level of resources and with low to high job demands.

We want to examine whether the assumed correlations between health inequality, working conditions at the training place and educational inequality are observable and how strong the correlations are. Furthermore, we assume that educational inequality is reflected in an unequal distribution of workloads (resulting from the working conditions at the training place) and unequal health opportunities. This could be caused by the social reproduction of life chances depending on education. By means of the categorial principal component analysis, we explore whether the health inequality, the workload (resulting from the working conditions at the training place) and the educational inequality can be explained by a dimension that depicts the social reproduction of life chances. Then, furthering Hillmert's (2014) work, who deals with the social reproduction of social status, one could speak of a social "inheritance" of life chances depending on education. The cluster analysis then allows for the identification of vulnerable groups on the basis of the results of the categorical principal components analysis.

3. Methodological approach

3.1 Sample

The survey data of trainees from ten different training occupations in Magdeburg, which were collected in 2015, are used to investigate the question. The occupations include the following: hairdresser, cook, baker, IT specialist, electrical technician, motor vehicle mechatronics technician, industrial business management assistant, wholesale trader and export merchant, trained retail salesman and bank clerk.

The selected training occupations are characterised by very different contract termination rates. Training as a hairdresser (61.7%) and cook (59.3%) are associated with a very high contract termination rate, while the occupations of bank clerk (8.8%) or industrial business management assistant (17.4%) are characterised by a very low contract termination rate (Kropp et al. 2014). The contract termination rate in Saxony-Anhalt has risen from 29.2% in 2010 to 31.9% in 2020, making it one of the highest in Germany (Kropp et al. 2014).

The survey was conducted in vocational schools. A total of 1124 trainees were participating in the training courses, of which 931 were surveyed by a written questionnaire that was distributed in class. A response rate of 82.8% was achieved. The descriptive results of the sample can be seen in Table 2.

Table 2: Description of the sample

Characteristics	Feature characteristics	Frequency	in %
Sex	male	578	62,1
	female	337	36,2
	not specified	16	1,7
Age	16 to 18	163	17,5
	19 to 20	313	33,6
	21 to 22	173	18,6
	23 and older	253	27,2
	not specified	29	3,1
Education	Hairdresser	104	11,2
	Cook	70	7,5
	Baker	41	4,4
	IT specialist	119	12,8
	Electrical technician	125	13,4
	Motor vehicle mechatronics technician	105	11,3
	Industrial business management assistant	40	4,3
	Wholesale trader and export merchant	113	12,1
	Trained retail salesman	153	16,4
	Bank clerk	61	6,6
Year of education	1st year of education	300	32,2
	2nd year of education	317	34,0
	3rd year of education	314	33,7
School-leaving qualification of the trainees	No school-leaving qualification / Special school	6	0,6
	Secondary school diploma	104	11,2
	Intermediate school-leaving certificate	534	57,4
	Abitur	249	26,7
	Others	11	1,2
	don't know / not specified	27	2,9

3.2 Operationalization

Table 3 gives an overview of the questions used for the survey about health, working conditions at the training place (job demands and resources) and educational inequality. The health status of the trainees has been recorded in a differentiated manner. Mental health was surveyed using the indicators for burnout. Physical health was recorded with two indicators that make physical stress at the workplace measurable. Social health was surveyed by means of an item that measures the social integration of the trainee in the company.

The working conditions at the training place were surveyed using measuring instruments that make it possible to measure the quality of the training conditions, the excessive demands, the job control and the social support (colleagues, trainers) of the trainees. The questions are linked to the explanatory model "Job Demands Resources Model" (Bakker & Demerouti 2007). The job demands are measured by the quality of the training conditions and the excessive demands. The job resources are made measurable on the basis of the perceived level of job control and social support.

Table 3: Operationalization

Theoretical concept	Item (Translated from German)	Cronbachs Alpha	Source
Burnout 1 = never to 5 = always	How often do you feel exhausted due to the work in the training company?	0,82	Based on the Next study (Simon et al 2005), specified to the topic of education.
	How often do you feel tired due to the work in the training company?		
	How often do you feel weak and prone to illness as a result of the work in the training company?		
Physical stress 1 = does not apply at all to 5 = applies completely	My work in my training company is physically demanding.	0,75	Own development
	I have to adopt a posture in my work in my training company that puts physical strain on me.		
Social health 1 = does not apply at all to 5 = applies completely	I feel comfortable and fulfilled in my job at my training company	not predictable	Own development

Theoretical concept	Item (Translated from German)	Cronbachs Alpha	Source
Quality of training conditions 1 = does not apply at all to 5 = applies completely	In the training company there are enough devices, tools and technical equipment for my training.	0,70	Own development
	In the training company, books, learning materials and media are available to me that are up to date.		
	The rooms in the training company where I work are in good condition.		
Excessive demand 1 = does not apply at all to 5 = applies completely	The responsibility imposed on me in my training company overwhelms me.	0,77	Following Satow (2013), specified on the topic of training.
	I am asked to do tasks in my training company which I am not yet able to do because of my level of training.		
	In my training at my training company, there are tasks that are too complicated for me.		
Job control	I get enough time to try out and practise new work tasks in the training company.	0,83	Following the COPSOQ study (Nübling et al. 2005), specified on the topic of training.
	I can independently plan, carry out and control the work in the training company.		
	I have an influence on what work is assigned to me in my training company.		
	I can decide for myself how to carry out the tasks assigned to me in my training company.		
	I can determine when I do the tasks assigned to me in my training company.		

Theoretical concept	Item (Translated from German)	Cronbachs Alpha	Source
Social support from colleagues 1 = does not apply at all to 5 = applies completely	My colleagues appreciate the value and results of my work in the training company. My colleagues express their opinions about my work to me. I receive appropriate support from my work colleagues in difficult situations.	0,71	Following the COPSOQ study (Nübling et al. 2005), specified on the topic of training
Social support from trainers 1 = does not apply at all to 5 = applies completely	My trainer appreciates the value and results of my work in the training company. My trainer expresses his opinion to me about my work in the training company. I receive appropriate support from my trainer in difficult situations in my training company.	0,83	Following the COPSOQ study (Nübling et al. 2005), specified on the topic of training
Grade point average	What grade did you get in German on your last school-leaving or graduation report? What maths grade did you have on your last leaving or final school report?	not calculable	Own development

Theoretical concept	Item (Translated from German)	Cronbachs Alpha	Source
School-leaving qualification			
No degree; Graduation from a special school; Secondary school diploma; Intermediate school-leaving certificate, Secondary technical school-leaving certificate or other intermediate school-leaving certificates; Higher education entrance qualification / Abitur, Advanced technical college entrance qualification; others	What is your highest school-leaving qualification? What is your mother's highest school-leaving qualification? What is your father's highest school-leaving qualification?	not calculable	Own development

The educational inequality was recorded on the basis of the trainees' school-leaving qualifications and their grades in German and mathematics. In addition, the school-leaving qualifications of the trainees' mothers and fathers were also recorded.

Schmitt (1996) pointed out in his study that measurement instruments that have a Cronbach's alpha value higher than 0.7 can be described as reliable. Overall, the measurement instruments meet the reliability requirements of Schmitt (1996) and are thus considered to be reliable.

3.3 Approach to statistical analysis

Categorical principal components analysis can reduce the selected data to a few relevant dimensions and make them representable (Linting et al. 2007; Meulman et al. 2002). In a coordinate system, the two dimensions represent the X and Y axes. The interviewees can be represented and described as points and the variables as axes in this two-dimensional space (Linting et al. 2007). The axes representing the variables can be of different lengths. The axes can also be situated at different angles in relation to the two dimensions and the other axes. Axes that are orthogonal to other axes or dimensions are not correlated with each other. Axes that are in a close relationship to other axes / dimensions indicate a correlative relationship. The cosine of the angle expresses the correlation (Linting et al. 2007).

Based on the correlative relationships, the dimensions can be described, taking into account the axes, in order to express relevant correlations in the data. The description of the latent dimensions makes it possible to answer the question of how the collected data can be explained and represented. Based on the research question of this article, the data is to be examined by means of the categorial principal components analysis, in order to determine how strong the correlations are between health inequality, working conditions at the training place and educational inequality. In addition, we will also consider to what extent the corresponding variables can be described and explained by means of the extracted dimensions of the categorial principal component analysis. We will also examine whether one dimension or several dimensions are suitable for describing and explaining health inequality, working conditions at the workplace and educational inequality.

In addition, the interviewed trainees can also be positioned in the two-dimensional space (Linting & van der Kooij 2012). The trainees are located on the two dimensions in the coordinate system based on their object values. Hierarchical cluster analysis is used to identify homogeneous groups of trainees. The extracted latent dimensions are used as the data basis for a hierarchical cluster analysis to identify homogeneous groups of trainees. This is because the latent dimensions represent the complex data in a reduced way.

4. Results

4.1 Descriptive statistics

Table 4 provides an overview of the correlation between the surveyed characteristics for the state of health, the working conditions at the training place (demands and resources) and the educational inequality among the trainees depending on the training occupation.

The health-related variables differ significantly between training occupations. For example, the experience of physical strain is particularly pronounced in training occupations for bakers, motor vehicle mechatronics technicians or hairdressers. The experience of physical strain is less pronounced in training occupations for bank clerks, IT specialists or industrial business management assistants.

Table 4: Descriptive statistics

EDUCATION		Baker	Hair-dresser	Cook	Motor vehicle mechatronics technician	Trained retail salesman	Electrical technician	Wholesale trader and export merchant	IT specialist	Industrial business management assistant	Bank clerk	Significance
Health												
Physical stresses (the higher the MV, the more stress)	MV	3,6	3,4	3,1	3,5	2,7	2,9	2,4	1,8	1,8	2,0	***
Burnout (the higher the MV, the more burnout)	MV	3,2	3,2	2,9	2,8	2,8	2,8	2,8	2,3	2,6	2,5	***
Social health (the higher the MV, the better the social health)	MV	3,5	3,7	3,4	3,5	3,4	3,2	3,4	3,7	3,5	3,9	**
Working conditions at the training place												
Quality of training conditions (the higher the MV, the better the conditions)	MV	3,6	3,9	3,5	3,6	3,8	3,7	3,9	4,2	4,2	4,1	***
Job control (the higher the MV, the more job control)	MV	2,8	3,0	2,5	2,4	2,9	2,6	3,0	3,3	3,2	3,4	***
Excessive demand (the higher the MV, the more excessive demand)	MV	1,9	1,9	2,1	2,2	2,0	2,2	2,1	2,1	2,0	1,9	***
Social support from colleagues (the higher the MV, the more support)	MV	4,0	3,9	3,7	3,8	3,8	3,8	3,9	4,0	4,2	4,3	***
Social support from trainers (the higher the MV, the more support)	MV	3,8	4,0	3,5	3,8	3,5	3,5	3,6	3,9	4,0	4,0	***
Educational inequality												
German grade	MV	2,9	2,8	3,0	2,9	2,6	2,8	2,3	2,4	2,2	2,3	***
Maths grade	MV	2,9	3,3	3,1	2,7	3,0	2,8	2,6	2,3	2,5	2,7	***
Proportion of trainees with Abitur	in %	0%	2%	12%	10%	17%	26%	36%	53%	47%	78%	***
Proportion of fathers with Abitur	in %	9%	13%	15%	26%	20%	23%	23%	28%	26%	23%	n. s.
Proportion of mothers with Abitur	in %	9%	9%	17%	32%	19%	24%	25%	29%	24%	25%	**

Significant differences were determined according to the scale level of the variables either with the F-test or with the Chi²-test.
*** p < 0.001; ** p < 0.01; n.s. = not significant; MV = mean value

The working conditions at the training place are partly connected to health-related conditions. This is because the apprenticeship occupations with a comparatively positive assessment of health also have partly better working conditions. For example, apprenticeship occupations for bank clerks, IT specialists or industrial business management assistants tend to have better assessed working conditions at the training place compared to the other apprenticeship occupations.

There is also a clear difference in the educational backgrounds held by trainees. For example, a higher proportion of trainees with an Abitur enrol in training for occupations such as bank clerk, IT specialist or industrial business management assistant. In addition, they have better grades in German and in mathematics compared to the other trainees, and their parents are also more likely to have a higher school-leaving qualification.

4.2 Results of the categorical principal components analysis

The two dimensions identified by the categorical principal components analysis can explain 45.1 % of the variance of the 13 variables.[1]

Table 5: Variance explained

Dimension	Eigenvalue	% of variance
1	3,7	28,7
2	2,1	16,4
Total	5,8	45,1

Table 6 gives an overview of the loadings of the variables in the two dimensions. The component loadings can be interpreted as correlations between the variables and the dimensions. High absolute values indicate a high correlation, while the sign indicates the direction of the correlation.

In dimension 1, high, negative loadings can be found for social health, the quality of the training conditions, the job control and the social support from work colleagues and trainers. Thus, high values for these variables are related to negative, low values in dimension 1. Physical stress, burnout and excessive demands are positively loaded with a high value in dimension 1. The experience of physical stress, excessive demands and burnout are thus related to a rather positive, high value in dimension 1. The health and work-related variables are thus very highly associated with each other and are closely related to dimension 1. Therefore, dimension 1 can be interpreted as a dimension for work-

---

[1] The results of the categorical principal components analysis do not differ with regard to their statements even if the calculation is carried out separately for the male and female gender.

related health inequalities. High, positive values in the dimension stand for high workloads due to excessive demands, low social support, little job control and poor training conditions which result in low social health, high physical stress and risk of burnout.

Table 6: Component loading

Component loadings	Dimension	
	1	2
Health		
Physical stress	0,50	-0,14
Burnout	0,65	0,01
Social health	-0,77	-0,19
Working conditions at the training place		
Working conditions:		
Training conditions	-0,65	-0,05
Excessive demand	0,67	0,25
Resources:		
Job control	-0,71	0,06
Social support from colleagues	-0,69	-0,07
Social support from trainers	-0,74	-0,12
Educational inequality		
German grade of the trainee	-0,08	0,47
Maths grade of the trainee	-0,11	0,43
School-leaving qualification of the trainee	-0,20	0,64
School-leaving qualification of the father	-0,11	0,74
School-leaving qualification of the mother	-0,09	0,79
Apprenticeship *	-0,30	0,31

Normalisation with variable principal.

*Additional passive variable

Dimension 2 has high, positive loadings from the school-leaving qualifications of the trainee and the parents. A high school-leaving qualification is positively related to dimension 2. Furthermore, high, positive loadings of the German and maths grade are found in the dimension. The grade "very good (=A)" is coded as "6" and therefore is regarded as a high numerical value of the grade variables, which is related to a high value in dimension 2. Accordingly, dimension 2 can be understood as a dimension for educational inequality. High values in dimension 2 are related to good grades and a high school-leaving qualification of the trainee and the parents, while a low value in the dimension stands for bad grades and a low school-leaving qualification of the trainee and their parents.

Dimension 1, which represents work-related health inequality, has an orthogonal and thus uncorrelated relationship with dimension 2, which depicts educational inequality. This can also be seen clearly in Figure 3. Consequently, a high level of education is not correlated with work-related health inequalities among trainees.[2]

In addition to the loadings, Figure 3 also shows the trainees as points in the biplot. The points can be projected into the space using the object values of the trainees, giving an indication of the expression in the dimensions. It can be seen from the figure that, with the exception of a few outliers around the axis cross, the trainees have spread out in a circle in the coordinate system.

Figure 3: Biplot (loadings and trainees shown)

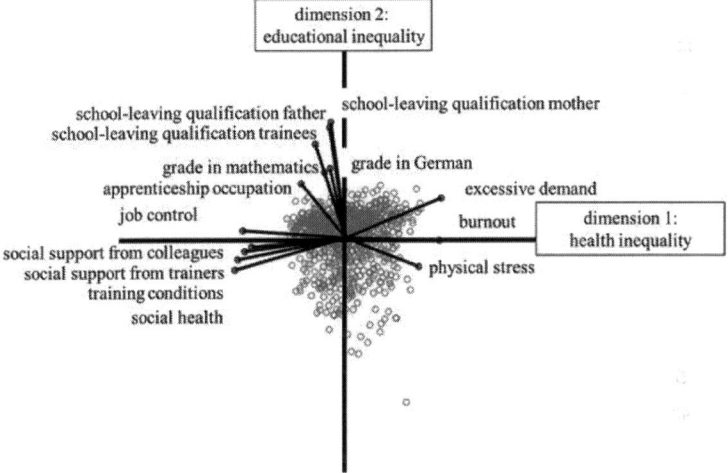

4.3 Results of the cluster analysis

The two dimensions of the categorical principal components analysis have been used as the data basis for the cluster analysis. The hierarchical clustering is done using the Ward method. Figure 4 shows the unification process by which the cluster solution with five groups was selected, because the heterogeneity jump would obviously increase significantly with a 4-cluster solution.

---

2   The result that the health-related indicators are not related to the education-related indicators has also been confirmed by means of regression analyses.

Figure 4: Cluster solution

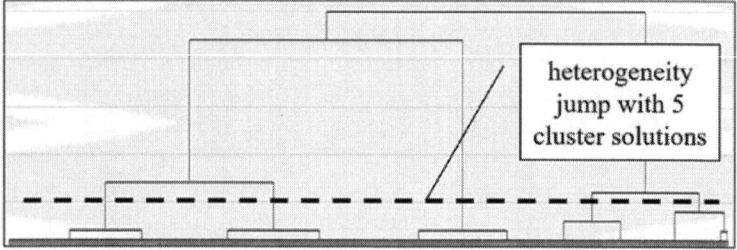

Figure 5 shows the clusters in relation to their object values in the two dimensions in the coordinate system. The representation makes it clear that there are five non-overlapping clusters. Table 7 presents selected statistical key figures depending on the cluster solution and thus makes it possible to describe the clusters. In addition to the variables considered so far, additional variables have also been used to describe the clusters.

Figure 5: Cluster solution

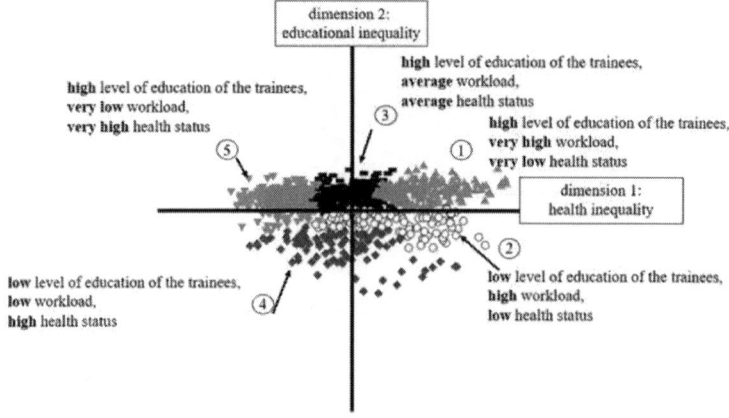

Clusters 2 and 4 are comprised of trainees who selected the training occupation of hairdresser, cook and baker more often than average. Educational resources tend to be below average in these training occupations. The trainees in the two clusters tend to achieve poor grades in German and mathematics. Furthermore, only a small proportion of the trainees from cluster 2 have an Abitur and none of the trainees in cluster 4 have an Abitur. The parents of the trainees in clusters 2 and 4 also have only a small proportion of Abitur. Differences can be seen between clusters 2 and 4 in the working conditions at the training place.

On average, the trainees in cluster 2 report comparatively worse working conditions at the training place than the trainees in cluster 4. The unfavourable working conditions at the training place are related to a higher risk of burnout, physical stress experience and low social health at the workplace. Compared to the trainees in cluster 4, the trainees in cluster 2 state that they develop a lower commitment to work, a higher willingness to drop out of training and they are less willing to work in their chosen training company.

Clusters 1 and 3 are comprised of trainees who selected the training occupation of motor vehicle mechatronics technician, electrical technician and wholesale trader and export merchant more often than average. The educational backgrounds in these clusters are average to above average. They tend to achieve average grades in German and mathematics. However, at least one in four of the trainees from clusters 1 and 3 has an Abitur. The parents of the trainees differ regarding their school-leaving qualifications. Every third father and every fifth mother in cluster 1 has an Abitur. In cluster 3, every fifth father and every third mother has an Abitur. Differences between cluster 1 and 3 are found in the working conditions at the training place. On average, the trainees in cluster 1 report comparatively worse working conditions at the training place than the trainees in cluster 3. The unfavourable working conditions at the training place are related to higher physical stress, risk of burnout and low social health at the workplace. Compared to the trainees in cluster 3, the trainees in cluster 1 state that they develop a lower commitment to work, have a higher willingness to drop out of training and that they work in their desired training company to a lesser extent.

Cluster 5 is comprised of trainees who selected the training occupation of IT specialist and bank clerk more often than average. Educational resources are high in this cluster. They tend to achieve above-average grades in German and mathematics. In addition, one in four of the trainees from cluster 5 has an Abitur. If we look at the clusters comparatively, we find that the trainees from cluster 5, compared to the other clusters, perceive the lowest level of physical stress, burnout risk and the highest level of social health at work. At the same time, the trainees in this cluster experience comparatively better working conditions at the training place. This is related to a higher level of work commitment, a lower willingness to drop out of the training and a high proportion working in the training company of their choice.

Table 7: Descriptive statistics for the cluster solution

	1: high level of education, poor working and health conditions	2: low level of education, rather poor health and working conditions	3: high level of education, average working and health conditions	4: low level of education, good health and working conditions	5: high level of education, very good health and working conditions	Total	
Training occupation							
Hairdresser	7%	21%	7%	24%	7%	11%	***
Cook	8%	14%	6%	13%	3%	8%	***
Baker	2%	13%	3%	7%	2%	4%	***
IT specialist	7%	0%	17%	9%	23%	13%	***
Electrical technician	19%	11%	13%	12%	9%	13%	*
Motor vehicle mechatronics technician	16%	13%	15%	8%	5%	11%	***
Industrial business management assistant	2%	0%	6%	3%	6%	4%	**
Wholesale trader and export merchant	16%	5%	11%	9%	15%	12%	*
Trained retail salesman	20%	22%	14%	13%	15%	16%	n.s.
Bank clerk	3%	1%	8%	1%	15%	7%	***
Health							
Physical stresses	3,3	3,4	2,6	2,8	2,0	2,7	***
Burnout	3,5	3,2	2,7	2,7	2,1	2,8	***
Social health	2,3	3,1	3,5	4,0	4,4	3,5	***

	1: high level of education, poor working and health conditions	2: low level of education, rather poor health and working conditions	3: high level of education, average working and health conditions	4: low level of education, good health and working conditions	5: high level of education, very good health and working conditions	Total	
Working conditions at the training place							
Work conditions:							
Quality of training conditions	3,2	3,5	3,8	4,0	4,5	3,8	***
Excessive demand	2,7	2,1	2,1	1,8	1,6	2,1	***
Resources:							
Job control	2,2	2,5	2,9	2,9	3,6	2,9	***
Social support from colleagues	3,1	3,6	3,9	4,1	4,5	3,9	***
Social support from trainers	2,7	3,4	3,8	4,0	4,5	3,7	***
Educational inequality							
German grade of the trainee	2,6	2,9	2,5	2,9	2,5	2,6	***
Maths grade of the trainee	2,8	3,1	2,7	3,2	2,5	2,8	***
Proportion of trainees with Abitur	29%	7%	25%	0%	27%	21%	***
Proportion of fathers with Abitur	31%	8%	24%	0%	30%	22%	***
Proportion of mothers with Abitur	18%	9%	38%	11%	39%	27%	***

	1: high level of education, poor working and health conditions	2: low level of education, rather poor health and working conditions	3: high level of education, average working and health conditions	4: low level of education, good health and working conditions	5: high level of education, very good health and working conditions	Total	
Descriptive variables							
Labour management	3,2	3,4	3,5	3,6	3,8	3,5	***
Desired company	8%	11%	19%	25%	32%	20%	***
Drop out of training	2,6	2,1	1,6	1,6	1,2	1,8	***
Proportion of women in the training occupation	35%	49%	31%	41%	37%	37%	*

Significant differences were determined according to the scale level of the variables either with the F-test or with the Chi²-test.

*** p < 0,001 ; ** p < 0,01 ; * p < 0,05 ; n.s. = not significant

Clusters 1, 3 and 5 have comparatively similar levels of educational resources but differ in terms of the estimated working conditions at the training place and the resulting social, physical and mental health. This is also shown by clusters 2 and 4, which draw attention to the fact that different working conditions at the training place and thus different social, physical and mental health levels are accessible with low educational resources. Accordingly, the working conditions at the training place are linked to health outcomes. It is noticeable from the results that trainees who do not work in their desired company struggle with unfavourable health and working conditions at the training place and as a result they think about dropping out of training.

5. Discussion of the results, limitations of the study and outlook

From our point of view, the results expand the existing state of knowledge. This is because there have been no studies so far that have examined the connection between health inequality, working conditions in training and educational inequality. The present study shows that the working conditions at the training place (job demands and resources) are closely related to health-related indicators. This is because the result of the categorical principal components analysis indicates that a dimension called work-related health inequality can explain both working condition indicators at the training place as well as health-related indicators. This result supports the previous research findings that occupational science approaches are suitable for explaining health-related indicators.

Health science approaches point to the importance of education in explaining social, mental and physical health. There are several studies which point out the correlation between health inequality and the level of education (Mielck 2012; Cutler & Lleras Munev 2006). In the present analysis, it becomes clear that the second dimension can be understood as educational inequality, which is independent and uncorrelated to the first dimension, work-related health inequality.

Furthermore, the component loadings show that the educational background of parents has an influence on the educational success of trainees. This is because the school-leaving qualifications of the parents are related to the school-leaving qualifications and grades of the trainees. Consequently, a social reproduction of educational qualifications can be seen well here.

However, educational qualifications do not wholly determine the health status of trainees. Trainees who have the opportunity to train in their company of choice, regardless of their level of education, have significantly better working conditions and access to better health conditions. They also show higher work commitment and are less likely to drop out of the vocational training.

Although the present study has limitations, these constraints can act as starting points for further, more in-depth studies. The study was only conducted for the Magdeburg area, a German city in Saxony-Anhalt with a population of just under 240,000. A further study might include both smaller and larger cities, as well as rural regions in other parts of Germany, to investigate whether there are different results due to the location of the training. The study is also limited because only 10 of almost 330 current training occupations have been considered. It would therefore be conceivable to extend the study to other training occupations. Perhaps then a connection between health inequality, working conditions and educational inequality could be proven.

References

Bakker, Arnold B.; Demerouti, Evangelia (2007): The Job Demands-Resources model: state of the art. In: Journal of Managerial Psychology 22, 3, pp. 309–328.

Becker, Rolf; Lauterbach, Wolfgang (2016): Bildung als Privileg: Erklärungen und Befunde zu den Ursachen der Bildungsungleichheit. Wiesbaden: Springer VS.

Bundesinstitut für Berufsbildung (BIBB) (ed.) (2022): Datenreport zum Berufsbildungsbericht 2022. Informationen und Analysen zur Entwicklung der beruflichen Bildung. Bonn.

Bundesinstitut für Berufsbildung (BIBB) (ed.) (2021): Verzeichnis der anerkannten Ausbildungsberufe 2021. Bonn.

Busemeyer, Marius R.; Trampusch, Christine (eds.) (2012): The Political Economy of Collective Skill Formation. Oxford: Oxford University Press.

Cutler, David; Lleras-Muney, Adriana (2006): Education and Health: Evaluating Theories and Evidence. In: National Bureau of Economic Research, 12352.

Duc, Barbara; Lamamra, Nadia (2022): Apprentices' health: Between prevention and socialization. In: Safety Science, 147, p. 1–10.

Gessler, Michael (2017): Educational Transfer as Transformation: A Case Study about the Emergence and Implementation of Dual Apprenticeship Structures in a German Automotive Transplant in the United States. In: Vocations and Learning 10, 1, pp. 71–99.

Granato, Mona; Ulrich, Joachim Gerd (2014): Soziale Ungleichheit beim Zugang in eine Berufsausbildung: Welche Bedeutung haben die Institutionen? In: Zeitschrift für Erziehungswissenschaften 17, S2, pp. 205–232.

Hillmert, Steffen (2010): Betriebliche Ausbildung und soziale Ungleichheit. In: Sozialer Fortschritt, 2010 59, 6–7, pp. 167–174.

Hillmert, Steffen (2011): Bildungszugang, Bildungskonsequenzen und soziale Ungleichheit im Lebenslauf. In: GWP – Gesellschaft. Wirtschaft. Politik, 2, pp. 177–190.

Hillmert, Steffen (2014): Bildung, Ausbildung und soziale Ungleichheiten im Lebenslauf. In: Zeitschrift für Erziehungswissenschaften 17, S2, pp. 73–94.

Kaminski, A.; Nauerth, A.; Pfefferle, P. I. (2008): Health status and health behaviour of apprentices in the first year of apprenticeship – first results of a survey in vocational training schools in Bielefeld. In: Gesundheitswesen 70, 1, pp. 38–46.

Kropp, P.; Danek, S.; Purz, S.; Dietrich, I. und Fritzsche, B. (2014): Die vorzeitige Lösung von Ausbildungsverträgen: eine Beschreibung vorzeitiger Lösungen in Sachsen-Anhalt und eine Auswertung von Bestandsdaten der IHK Halle-Dessau, No 201413, IAB-Forschungsbericht, Institut für Arbeitsmarkt- und Berufsforschung (IAB) (ed.). (unpubl.).

Li, Junmin; Pilz, Matthias (2021): International transfer of vocational education and training: a literature review. In: Journal of Vocational Education & Training, 1, pp. 1–34.

Linting, Mariëlle; Meulman, Jacqueline J.; Groenen, Patrick J. F.; van der Kooij, Anita J. (2007): Nonlinear principal components analysis: introduction and application. In: Psychological Methods 12, 3, pp. 336–358.

Linting, Mariëlle; van der Kooij, Anita (2012): Nonlinear principal components analysis with CATPCA: a tutorial. In: Journal of Personality Assessment 94, 1, pp. 12–25.

Maaz, Kai; Neumann, Marko; Baumert, Jürgen (ed.) (2014): Herkunft und Bildungserfolg von der frühen Kindheit bis ins Erwachsenenalter. Forschungsstand und Interventionsmöglichkeiten aus interdisziplinärer Perspektive. Wiesbaden: Springer VS (Zeitschrift für Erziehungswissenschaft. Sonderheft. 24).

Meulman, J. J.; van der Kooij, A. J.; Babinec, A. (2002): New Features of Categorical Principal Components Analysis for Complicated Data Sets, Including Data Mining. In: Classification, Automation, and New Media. Heidelberg. Berlin: Springer VS, pp. 207–217.

Mielck, Andreas (2012): Soziale Ungleichheit und Gesundheit. Empirische Belege für die zentrale Rolle der schulischen und beruflichen Bildung. In: Brähler, Elmar (ed.); Kiess, Johannes (ed.); Schubert, Charlotte (ed.); Kiess, Wieland (ed.): Gesund und gebildet. Voraussetzungen für eine moderne Gesellschaft. Göttingen: Vandenhoeck & Ruprecht, pp. 129–145.

Ministerium für Arbeit, Soziales und Integration des Landes Sachsen-Anhalt (MS) (ed.) (2021): Jahresmonitor Berufsbildung Sachsen-Anhalt 2020.

Nübling, M., Stößel, U., Hasselhorn, H. M., Michaelis, M., & Hofmann, F. (2005): Methoden zur Erfassung psychischer Belastungen. Erprobung eines Messinstrumentes (COPSOQ). Bremerhaven: Wirtschaftsverlag NW.

Protsch, Paula; Solga, Heike (2016): The social stratification of the German VET system. In: Journal of Education and Work 29, 6, pp. 637–661.

Rohrbach-Schmidt, Daniela; Uhly, Alexandra (2015): Determinanten vorzeitiger Lösungen von Ausbildungsverträgen und berufliche Segmentierung im dualen System. Eine Mehrebenenanalyse auf Basis der Berufsbildungsstatistik. In: KZfSS 67, 1, pp. 105–135.

Satow, L. (2012): SCI. Stress- und Coping-Inventar [Verfahrensdokumentation, Fragebogen, Skalendokumentation und Beispielprofile]. In Leibniz-Institut für Psychologische Information und Dokumentation (ZPID) (ed.). Trier: ZPID.

Schmitt, Neal (1996): Uses and abuses of coefficient alpha. In: Psychological Assessment 8, 4, pp. 350–353.

Schaufeli, Wilmar B.; Taris, Toon W. (2014): A Critical Review of the Job Demands-Resources Model: Implications for Improving Work and Health. In: Wilmar B. Schaufeli (ed.): Bridging Occupational, Organizational and Public Health. Wiesbaden: Springer VS, Dordrecht, pp. 43–68.

Schlicht, Raphaela (2011): Das Konzept soziale Bildungsungleichheit. In: Schlicht-Schmälzle, Raphaela (ed.): Determinanten der Bildungsungleichheit. Die Leistungsfähigkeit von Bildungssystemen im Vergleich der deutschen Bundesländer. Wiesbaden: Springer VS., pp. 35–44.

Siegrist, Johannes; Dragano, Nico (2020): Arbeit und Beschäftigung als Determinanten ungleicher Gesundheit. In: Jungbauer-Gans, Monika; Kriwy, Peter (ed.): Handbuch Gesundheitssoziologie. Wiesbaden: Springer VS, pp. 235–251.

Simon, M., Tackenberg, P., Hasselhorn, H. M., Kümmerling, A., Büscher, A., & Müller, B. H. (2005). Auswertung der ersten Befragung der NEXT-Studie in Deutschland. Universität Wuppertal.

Statistisches Bundesamt (2022): Bildung und Kultur, Berufliche Bildung, Fachserie 11, Reihe 3, Berichtszeitraum 2021.

World Health Organization (2010): A conceptual framework for action on the social determinants of health. Debates, policy & practice, case studies. Geneva: World Health Organization (Social determinants of health discussion paper, 2).

Part III: Social inequality and change in (urban) space

Gentrification as a self-producing and self-reenforcing process on the macro, meso and micro level

Jens S. Dangschat

1. Introduction

Gentrification is an almost worldwide phenomenon – and is a widely used buzz-word. Definitions of 'gentrification' widened parallel to its territorial spread – within urban agglomerations and outward in rural areas or as a global phenomenon as well as in capitals of the Global South. While the first debate in the 1980s was about whether gentrification is driven by the demand or supply side, the second in the 1990s was about whether rent gap or value gap[1] might be the better explanation of capital re-investment in inner Western cities. Recent debate is split in either shedding light on gentrification as a global (planetary) phenomenon or discussing whether the local level is relevant.

One other perspective was mentioned by Hamnett, who described the battlefield in the 1990s as stage of patricide during a personal talk: Gentrification as process (described by younger geographers and urban sociologists) contradicts present theories and empirical evidence of residential segregation by class, status or income (as argued by their teachers), because affluent people are expelled by lower income groups as they dislike those as neighbours and fear depreciation of the homes. Moreover, processes had been centrifugal in traditional

---

1 The *rent gap-position* describes the gap between the existing ground rent and the potential ground rent that can be achieved in the existing property, while *value gap* point to the difference of returns between a rental property and the potential profit after conversion into owner-occupied properties.

findings on segregation, while gentrification was directed towards the historic centres.

Recent debate neglects some aspects of the recent past. First: The link to the bulk of residential segregation research is low, and only economic or regulation aspects are highlighted, while cultural aspects of milieu and life-style as well as aesthetic aspects are widely neglected (Dangschat & Blasius 1994). Second: Gentrification theory must explain, why here and now driven by whom gentrification takes place (Hamnett 1991: 33), i.e. how specific places are produced and reproduced. Third, the dispute whether gentrification is a planetary *or* local phenomenon neglects the idea of glocalisation (Robertson 1995): How are global trends and narratives trickling down to urban places and what is the role of policy transfer? Fourth: Gentrification research is – despite of the theoretical arenas – predominately descriptive and based on case studies, with different political contexts, relative position within the agglomeration landscape, housing markets etc., but is hardly orientated at micro-meso-macro theories of the production and reproduction of space and place.

This article firstly describes – in a nutshell – the development of gentrification research in Germany and Austria. Secondly, it intervenes in the multi-layered 'either-or-debates' how gentrification is to be classified. Thirdly, I suggest a macro-meso-micro model based on Bourdieu's thoughts of the perpetuation of class hierarchies (as stable self-reinventing power order) and the production of space and allocation of space as one of the most subtle forms of power over people (Bourdieu 1986, 1990, 1998).

2. Gentrification in German-speaking countries

Gentrification research started in German-speaking countries a little bit belated in the late 1980s. Based on students' practical courses at Hamburg University, the first rather descriptive but also hypothesis testing study of gentrification in three neighbourhoods in Hamburg was published (Dangschat/Friedrichs 1988), followed by the first theoretical article (Dangschat 1988). First evidence and theoretical thoughts about the spreading of gentrification processes outside the Anglo-Saxon world were internationally discussed in the Utrecht conference in 1990 (Dangschat 1991b). Further quantitative and qualitative empirical research about Hamburg was conducted throughout the 1990s (Alisch/zum Felde 1990; Dangschat/Alisch 1995), later on a habilitation treatise about a broader theoretical understanding of gentrification (Dangschat 1991a) and a dissertation about the role of women in gentrification (Alisch 1993) were finished, followed by some master theses of students of sociology.

With the changeover of Jürgen Friedrichs to Cologne University, the interest in gentrification analysis shifted to Cologne (cf. a summary of research about the city: Friedrichs/Blasius 2016). The first conference on gentrification was held in 1999 in Cologne, collecting further research about different aspects of gentrification throughout Germany (Blasius/Dangschat 1990). Blasius (1993) published his PhD thesis about gentrification in a quarter of Cologne and in 1996, Friedrichs and Kecskes edited a second state-of-the-art of gentrification book for Germany; further state-of-the art reports were following by Glatter (2006); Blasius (2008), Eckardt (2018), Üblacker (2018) and Glatter/Mießner (2022b, c). The latter try to define a phase model of gentrification in Germany, but they indeed are following the documented research interest, which has much to do with the working town of the main researchers.

During Häußermann's professorship at Humboldt University in Berlin a new hotspot of research was established, which spread over the former GDR cities parallel to increasing rent and value gaps pushed by the transition from state-controlled housing and ground to market processes particularly in Berlin (Helbrecht 2016) and Leipzig (Haase/Rink 2016); for an early overview of gentrification research in Eastern Germany, cf. Bernt et al. (2010). Holm was taking the lead in these debates from a neo-Marxist position and wrote some 'gentrification'-chapters in readers (Holm 2012, 2014, 2020).

With further delay, gentrification research started in Vienna in Austria. Huber (2013) compared processes in Vienna with those in Chicago and Mexico City, while Kadi/Verlic (2019) – beside empirical research – shed more light on the role of politics and planning in a housing market that had long been strongly regulated. Kadi/Matznetter (2022) date back their gentrification research about Austria's capital to the mid-18th century, which is problematic insofar as they interpreted war damages and eviction of Jewish people as phases of modified gentrification processes. In a more theoretically based article, Kadi (2019) reflects what really is done worldwide when analysing gentrification processes within rather different contexts.

Recent debate in Germany revolves on the one hand – following Hackworth/Smith's (2001) argument for the third wave of capitalisation and Christophers' (2012) theses about the role of the growing of financial sectors share of profits – around commodification and financialisation of housing (Belina 2022, Janoschka 2022). On the other hand, authors discuss – impacted by Lees et al. (2008, 2016) – whether gentrification should be discerned as a local/particular or global/universal process (Bernt 2016, 2020).

3. Gentrification: More than an either – or!

Definitions of gentrification are either very narrow and therefore predestined for developing a theoretical definition like "gentrification is the replacement of lower strata in favour of higher strata within a neighbourhood" (Friedrichs

1996:14) or they include all kinds of upgrading and displacement everywhere throughout the world (Lees et al. 2016).

The definition used here follows Hamnett (1991: 32), who stated that gentrification is a "physical, economic, social and cultural phenomenon", an "invasion by middle-class people in former working-class districts or multi-occupied 'twilight areas'" and the "replacement and displacement of many of the original occupants." Other aspects are "physical renovation and rehabilitation", which causes "significant price appreciation" and commonly it "involves a degree of tenure transformation from renting to owning". Referring to the data of the Hamburg study I would add, that second and third waves of gentrification can also happen in quarters first built in the period of promoterism and/or art nouveau based on longer periods of incumbent upgrading (Dangschat/Alisch 1995) and are driven by tenure change to owner occupation. Displacement then is among middle class (middle-aged families of, say, teachers have to leave in favour of established managers who are double-income-no-kids households).

Empirical research predominantly is about gentrifying neighbourhoods and their immediate actors. Mainly they are classified as 'pioneers' and 'gentrifiers', while "the others" are named differently: old people, families, better-off or relatively non-specific middle classes. Educational level, income, household type and age are mainly used for classification. Only Alisch/Dangschat (1996) – following the life-course orientation – used shifts between the types while living in place, which is important concerning the citizens' appraisal for and support of changes of their neighbourhood. Here, not only pioneers are fighting further gentrification, but also the early gentrifiers themselves resist against the ultra-gentrifiers throughout the second wave of gentrification (tenure change to owner occupation) in their neighbourhoods.

Friedrichs (1996: 40) criticised economic and social process analysis within case studies due to their circular logic, as gentrification (as upgrading of the housing stock and increase of household income) is only researched in places, which are already labelled as gentrified areas without any control in other places.

Other work is about entire cities (to describe phases of gentrification and/or decline; cf. Kadi/Matznetter 2022; Reick 2018). This is problematic as gentrification is a spot-wise process along street sections (Berry 1985), where rent and value gaps are wide enough and can be closed. Already within bigger neighbourhoods under research the risk of grasping different stages of gentrification processes is real, which is blurring the analysis (Blasius 1993). This might be one of the reasons, why gentrification theory often is marked as chaotic (Beauregard 1986) or as failing (Ghertner 2015).

Gentrification research is a multi-faceted battlefield – among researchers, local politicians, within public administrations and between the established and the grassroots level. Gentrification as a buzzword is widely used in popular books, journals and newspapers. In this case it describes the clashes between traditional left- and right-wing positions. In cities, where gentrification is obvious, local administration is mostly classified as neo-liberal, overlooking however that departments of the cities might be in a very opposite position: economic departments support the processes of re-investment, upgrading and urban renaissance (economy, finance, industrial modernisation, technology (smart city), high-brow culture, tourism, etc.) or due to (foreign direct) investments and growing tax revenues for the urban agglomeration, while 'social' departments of cities (social housing, social policy, experimental culture, etc.) are criticising re- and displacement, loss of affordable housing and life-stylisation of the 'places to be'.

At first, debate was between neo-positivist geographers who underpin the impact of new demand for inner-city housing and aspects of culture and consumption (Blasius 1993, 2004; Boterman/van Gent 2023; Clay 1979; Dangschat, 1990, 1991b; Dangschat/Alisch 1995; Hamnett 2003; Lees 2008; Ley 1980, 1986, 1996; Rose 1984; Zukin 1982, 1987) and neo-marxist supply-side promotors (Badcock 1989; Buzek/Mießner 2022; Clark 1987, 1991; Marcuse 1986; Smith 1979, 1984, 1986, 1987, 1991, 1996). Hamnett (1984, 1991), Hamnett/Randolph (1986), Beauregard (1984) and Blasius et al. (2016) tried to bridge the first ‚either-or' debate arguing that demand and supply need one another. In his later works, Smith (1991, 1996) accepted arguments 'from the other side', but classified them always as secondary.

Second, there was battle within the supply-side fraction whether the rent gap or the value gap should be a better reference for analysis. First fractionation had been along the North America-Europe divide, as housing markets and particularly local policies are rather different. Clark (1987, 1991) was among the first who showed for Stockholm that both processes work hand in hand and need one another like a ping-pong, while Bernt (2022) pointed to the restrictions of the rent gap hypothesis for German cases. Smith (1991) insisted on the dominance of rent gap theories, while in cases of Western European cities value gap approaches are preferred.

Instead of following this debate, the understanding of gentrification was enlarged to all kinds of upgrading, neighbourhood change and changes of the mix of the residents throughout the world (for a critical overview, cf. Maloutas 2012). This development was partly due to the geographical enlargement of the processes of supply and demand in a multi-nested center-periphery movement (trickling down in global hierarchies of agglomerations and within them from the inner cities to the periphery and into attractive rural sites) (Phillips 1993, Stockdale 2010).

Hackworth/Smith (2001) and Smith (2002) opened a debate about new kinds of financial capital driven waves of (global) gentrification. They opened a third, recently ongoing battlefield about whether gentrification must be seen as a third wave, driven by new global processes of commodification and financialisation (Belina 2022; Janoschka 2022), followed by an interpretation of gentrification as 'planetary' (Lees et al. 2016; Slater 2017). Aalbers (2019) added a fifth wave by pointing to the new 'growth coalitions' between states and finance-led capitalism leading to a state-sponsored gentrification. Similar results were found already in the 1990s in Hamburg, as public subsidies for renovation had restricted the increase of rental fees for only eight to ten years and therefore are serving as a stepstone for further gentrification (Alisch/Dangschat 1998).

Ghertner (2015) – reflecting revitalisation in post-communist cities and big agglomerations in the Global South – proposed to lay to bed gentrification theories like other theories of the 20th century from the Global North. This debate was paralleled by the interpretation that both foreign-direct investment interests and scientific research from the Global North constitute a new form of colonialization (Atkinson/Bridge 2005; Ghertner 2015); this stream is strongly impacted by the post-colonialization debate in social sciences (Elam 2019) (for a critique at the 'postcolonial attacks' cf. Wyly 2019).

Within this debate again scepticism was raised, whether macro level aspects are more important than local/regional conditions to understand and explain different kinds of neighbourhood change (Berndt 2016, 2020) – for an overview about the impact of different local contexts cf. Atkinson/Bridge (2005) and Maloutas (2012). And again: it is not an 'either or' but referring to Robertson (1995) it needs the theorization and case study work to shed more light on the different drivers of gentrification and their interaction on different levels.

4. Gentrification as production and reproduction of places/spaces

To develop a gentrification theory beyond the different 'either-ors', you need to explain the *production of places and spaces* by narratives about inspiring neighbourhoods, (global) investments, dominant political cultures, embedded in different technological, economic, social and cultural transitions as a context of the macro level (global, macro-regional, national). The meso level is about cities/urban agglomerations and particularly those areas under pressure of gentrification. Again, political cultures, structures and regulations are important as well as the economic position (labour market, amenity for foreign direct investments), housing market and housing stock, segregation patterns and the dominant discourse about urban regeneration. On the micro level the citizens are relevant; they are supplied with resources and constraints, internalise their habitus form and have some routines of activities.

These logics must be related to the *reproduction of places and spaces*. According to Lefebvre (1991), place is a product of the dynamic between three elements:

1. everyday practices and perceptions of people (*'spatial practice'*), which sitting tenants and newcomers follow within their scope of action and attach a feeling of home to them;
2. cognitive concepts and theories of space (*'representational space'*) describe the way people 'read' and understand signs and symbols of gentrification processes;
3. while the first two aspects complement one another for average persons, the *'representations of space'* are spatial imaginations of professionals (scientists, policy makers, spatial planners and media) about vibrant places vs. displacement of people and loss of affordable housing.

Representations of space are relevant as well for self-reflections of scientists (i. e. the normative background in the sequence of dominant theories and for a critical reflection of the politicisation and political use of the term 'gentrification'). Thus, the publications and talks of 'guru' Florida can be seen as guiding principle for urban stakeholders to intensify the city-to-city competition by attracting the 'creative class' (Florida 2002). In his books he explores the factors that shape the 'quality of place and people' in changing cities by 'the 3ts': talent, technology, tolerance. Sassen (1991) – maybe not at all deliberately – supported with her 'global city' thesis, as bigger agglomerations pushed forward to achieve the 'champion league' of global hubs.

4.1 A macro-meso-micro model of societal space

The German economist Läpple (1991) developed a macro-meso-micro model of societal space. He highlighted five distinct aspects, where macro-level frames the range of diversification of the meso level, which in turn defines the setting for local human activities on the micro level. Here, his general societal space model is adopted for the analysis of multi-level gentrification aspects (cf. Fig. 1). Läpple's model is not restricted to the description and classification of concrete places, but admits that these are embedded in power structures of the macro level (cf. as well Massey 1993).

Figure 1: Macro-Micro-Meso Model of Societal Space

Macro-level	• *power structure* of global / national regulation (neo-liberal vs. welfare state orientation, role of finance and housing sector), global economy, finance system, societal macro trends, etc.
Meso-level	• *physical space* as built and 'natural' environment (buildings, infra-structure, housing, accessibility, landscape, etc.) • *societal space* (social structures, networks, local political culture, etc.) • *symbolic space* (cognition of the socio-spatial setting), both as individual cognition and as collective memory
Micro-level	• *local interaction* of people within the neighbourhood (supporting vs. fighting gentrification)

Own figure, after Läpple (1991)

The levels are enriched with power structures, forms of regulation, symbols and the interaction of people. However, this model of ‚societal space' disregards that people dispose of resources and constraints in different ways to cope with different kinds of restrictions of their social and/or physical environment. Therefore, Läpple's model needs to integrate an additional level for the societal inequalities (power, interests, capacity).

4.2 Bourdieu's class theory and his formula of structure-habitus-practice reproduction of social inequalities

According to Bourdieu (1986), class structure is determined by different amounts of 'capitals':

- *economic capital* – the amount of money, like income and wealth,
- *cultural capital* – formal educational attainments as 'institutionalised cultural capital', but as well the ownership of cultural goods (books, paintings, musical instruments) as 'objectified cultural capital'; and knowledge about consuming, using and acting with (particularly high-brow) cultural goods as 'incorporated cultural capital',
- *social capital* – social networks, ties and dealings,
- *symbolic capital* – reputation and prestige as much as the competence to act properly in different socio-spatial settings.

Beside the amount of these forms of capital, the ability to transfer one capital into another in different societal fields is important – this could be the housing market but as well how one acts in public places or fights against exclusion and tenure transformation from renting to owning (Dangschat 1990).

Bourdieu (1990) also considered how the traditional upper classes remain in their powerful positions. For him, social inequalities consist of three levels:

i) *structure* (the above-mentioned capital forms in their extent and flexibility to transfer them into one another), ii) *habitus* (set of values, aims, attitudes and tastes) and iii) *practice* (measurable activities).

According to Bourdieu the amount and structure of capital(s) is framing, but not determining the range of habitus forms and orientation, which in turn mitigates the variety of the potential of actions in practice. Moreover, the way of activities, which normally are in line with the habitus, can alter the values system slightly by interest coalitions, economic, societal and cultural change or social control of peers. Changes of habitus in turn might have an impact on different capital forms and/or their amount (as impact of consumption styles, time to invest in cultural consumption and learning at the expense of economic capital), as much as the transfer strategies.

Practice describes all activities by people – whether as common citizen, or in interest groups like NGOs, local initiatives and associations. Individuals are discerned in this model by the Bourdiesian classes, which frame the variety of habitus contents of people, which in turn enable different kinds of activities.

All stakeholders within gentrification processes (like policy-makers, urban planners, CEOs, investors, etc.) have to be considered as individuals as well. They are acting on the macro and meso level, but are nested within their respective organisational structure and value-system. Therefore, their capacities are restricted by their role. Their individual habitus must reflect the normative context of their organisation, and their room of manoeuvre should respect the organisational guide rails (Mayntz/Scharpf 1985).

By his structure-habitus-practice formula, Bourdieu (1990) illustrates, why and how societal power and mainstream value systems are relative stable and underpin interests of traditional upper classes for their perpetuation. To understand why gentrification processes are enforced almost worldwide under specific conditions, the Bourdiesian logic of the structure-habitus-practice relations can be made fruitful for the analysis of the driving forces of the production and reproduction of place and space.

4.3 A macro-meso-micro model of place and space its structures, habitus forms and practices

In earlier articles I claimed, that macro, meso, micro levels of place and space all have a *structure* like the organisation of governments and institutions, regulation of markets (finance capital, tax system, housing and urban renewal) on macro and meso level, and economic as much as cultural resources and constraints on the micro level (Dangschat 1990, 2007). Moreover, all levels are provided by a certain *habitus* (i. e. interest, conventions, aims and values of institutions and stakeholders on the macro and meso levels and of individuals on the micro level) embedded in traditional routines and habits which open the field for a spectrum of practice (among them also un-intended side-effects like rebound).

The idea to assign habitus forms to places dates back to Roman times and is widely used in architectural discourses as 'genius loci'. Tuan (1977) used the terms 'spirit' or a 'personality of places', while Cresswell (2013) named it 'sense of place' and I called it – basing on the ideas of Bourdieu (1998) – 'habitus of place' as an incorporation and inscription into bodies and places (Dangschat 2009). A habitus of place continues to exist, even if other people are active or at another place in time. Nevertheless, habitus is not fixed but has a kind of permanency due to the aggregated collective memory.

As *structures* define the relation of resources and constraints, *habitus* defines the values, aims and goals, which translates into day-to-day *practice*, which are on the *macro* and *meso level* acts, rules, decrees, and programmes of governments, investments of the finance sector (Hackworth/Smith 2001), and narratives about successful economic competition (Florida 2002), cultural and architectural highlights and urban revitalisation lighthouse projects (cf. Fig. 2).

Figure 2: Macro-meso-micro model of the production and reproduction of environments for older people

Macro (global, national)	Structure	national & regional government / markets (health, care, housing, transport, technologies, etc.) / macro trends (globalisation, digitalisation, climate change regulations, etc.)
	Habitus	self-understanding of political role (welfare state vs. workforce state) global / national / regional norms of capital investment in housing
	Practice	governance (laws, regulations, budget preferences, pension systems) / national & regional investments / age regimes / research programmes
Meso (regional, local neighbourhood)	Structure	regional & local governance / spatial planning / housing structure / infrastructure & service structure / economic structure (labour market) / social structure
	Habitus of Place	regional / local norms of housing policies, urban renewal & capital investment
	Practice	local governance (regulations, budget preferences) / investments in housing, renewal / local culture / social networks / adaption strategies of homes and public places
Micro (individual)	Structure	economic, social & cultural capital / resources & constraints
	Habitus	attitudes towards housing needs / attitudes towards upgrading and displacement
	Practice	demand on housing market

Own figure

Habitus and *practice* of global finance capital and political regulation circulate among countries at the macro level. Therefore, it needs more analysis, how policy and investment logic transfers really take place on the macro level from North America to Continental Europe, from traditional market economies to post-socialist cities, from Global North to Global South (Dolowitz/March 2000) and how they are trickling down to agglomerations, cities and neighbourhoods (Evans/Davies 1999). The urban-policy-mobilities view (McCann/Ward 2012) points to the fact, that administrative boundaries become more permeable in favour of an interplay of scales (Brenner 1998). Political, financial and technological practices open up new spaces, as territories are de-coded and re-coded ('assemblage', Deleuze/Guattari 1992). *Habitus* and *practice* on the macro level frame the structures and habitus on the meso level (of cities and neighbourhoods), leading to specific forms of practice in regional/local housing markets.

The *meso level* is the centre of interest in case studies. Government structures and politics of the past frame governance styles as result of the competition of different habitus forms of the departments of the cities (growth coalition vs. social inclusion). Neighbourhood changes are described by characteristics of old and new residents and/or housing stock changes, but little is said about the habitus of these places. Only in accordance with the fight against a more intense gentrification reflecting 'right to the city' (Mayer 2012) aims and value orientation (habitus) of the networks are mentioned and the forms of protest (practice) are described (investing their social and cultural capital in that field against economic capital of the investors and landlords, Dangschat 1990).

The *micro level* is about the citizens' cognition of the local/regional power structures, investments, policies and housing market, social composition of the neighbours and visitors as much as quality of public space. They consider particular capital forms of the old and new tenants, local political cultures and their own market position. Here, the individual housing career is important for identifying with the place. Those people whose own social status went up parallel to economic upgrading of the neighbourhood, like their (gentrified) neighbourhoods the most (Alisch/Dangschat 1996).

This view of distinction between structure, habitus and practice of place and space production and reproduction on different levels is important for shedding more light on the role of institutions as being socially structured. As only the sheer existence of institutions often is part of the operationalisation of 'neighbourhood effects', merely isolated aspects of institutions like policies, programmes, practices, norms or conventions are to be considered, which are setting social roles and behaviour.

5. Conclusion

Beauregard's (1986) valuation that gentrification is a chaotic concept seems to be still suitable – maybe it is real the more the number of battlefields increased. Therefore, Hamnett's (1991) interpretation of gentrification debate like the tale of 'the blind men and the elephant' is still relevant, as even more blind men and women are in the field, but the elephant is still not fully discovered. These 'blind men' should work more hand in hand and not attack one another at the frontiers.

If contemporary gentrification is, on the one hand, a global process with generalizations and on the other contextually framed, then it needs to overcome the either-or-debate. This endeavour needs to respect each others' works, which is concentrated at macro and meso level. Here the concepts of policy transfer and urban-policy-mobility are relevant for a more detailed and evidence-based analysis of the global third, fourth and fifth waves. As these analyses so far are very general, it is important to consider the structures of governments and finance capital, the different habitus forms in conflict and the resulting practices thereof.

Same holds for the meso level of the cities and the respective neighbourhoods and housing market segments, which is to some degree done in case studies. However, the 'Bourdiesian trinity' of structure, habitus and practice on both levels might help to pattern the research and to define categories for comparative analysis.

Interestingly enough, the micro level is more or less absent looking through the glasses of sociological inequality research and theories of action. Empirical work on local citizens is mainly based on structural indicators of SES, age and household type, only randomly on life-style or milieu categories or by insufficient dummies like journalists, gay people, Yuppies etc. Again, different habitus forms in conflict are relevant to understand the 'habitus of place' and the respective activities.

Using the suggested model, the methodological and empirical challenges of multi-level analysis need to be overcome. However, it might be used as a blueprint for different kinds of gentrification analysis, helping to bring some insights and order in the chaotic concept of gentrification.

References

Aalbers, Manuel B. (2019): Introduction to the Forum: From Third to Fifth-wave Gentrification. In: Tijdschrift voor Economische en Sociale Geografie 110, 1, pp. 1–11.

Alisch, Monika (1993): Frauen und Gentrification. Der Einfluß von Frauen auf die Konkurrenz um den innerstädtischen Wohnraum. Wiesbaden: Deutscher Universitätsverlag.

Alisch, Monika/Dangschat, Jens S. (1996): Die Akteure der Gentrifizierung und ihre „Karrieren". In: Friedrichs/Kecskes (eds.) (1996), pp. 95–129.

Alisch, Monika/Dangschat, Jens S. (1998): Armut und soziale Integration. Strategien sozialer Stadtentwicklung und lokaler Nachhaltigkeit. Opladen: Leske & Budrich.

Alisch, Monika/zum Felde, Wolfgang (1990): „Das gute Wohngefühl ist weg!" – Wahrnehmung, Bewertungen und Reaktionen von Bewohnern im Vorfeld der Verdrängung. In: Blasius/Dangschat (eds.) (1990), pp. 277–300.

Atkinson, Rowland/Bridge, Gary (eds.) (2005): Gentrification in a Global Context. The New Urban Colonialism. London: Routledge.

Badcock, Blair (1989): An Australian View of the Rent-Gap Hypothesis. In: Annals of the Association of American Geographers, 79, pp. 125–145.

Beauregard, Robert A. (1984): Structure, agency, and urban redevelopment. In Smith, Michael P. (ed.): Cities in Transformation. Urban Affairs Annual Reviews, 26, pp. 51–72.

Beauregard, Robert A. (1986): The chaos and complexity of gentrification. In: Smith/Williams (eds.) (1986), pp. 35–55.

Belina, Bernd (2022): Gentrifizierung und Finanzialisierung. In: Glatter/Mießner (eds.) (2022a), pp. 57–71.

Bernt, Matthias (2016): Very particular, or rather universal? Gentrification through the lenses of Ghertner and López-Morales. In: Cities 20, 4, pp. 637–644.

Bernt, Matthias (2020): Gentrifizierung zwischen Universalismus und Partikularismus. In: Breckner, Ingrid/Göschel, Albrecht/Matthiesen, Ulf (eds.): Stadtsoziologie und Stadtentwicklung. Handbuch für Wissenschaft und Praxis. Baden-Baden: Nomos, pp. 403–414.

Bernt, Matthias (2022): Die Grenzen der rent gap-Theorie. In: Glatter/Mießner (eds.) (2022a), pp. 91–106.

Bernt, Matthias/Holm Andrzej/Rink, Dieter (2010): Gentrificationforschung in Ostdeutschland: konzeptionelle Probleme und Forschungslücken: In: Berichte zur deutschen Landeskunde 84, 2, pp. 185–203.

Berry, Brian J.L. (1985): Islands of Renewal in Seas of Decay. In: Peterson, Paul E. (ed.): The New Urban Reality. Washington, DC: Brookings, pp. 69–96.

Blasius, Jörg (1993): Gentrification und Lebensstile. Wiesbaden. Deutscher Universitätsverlag.

Blasius, Jörg (2004): Gentrification und die Verdrängung der Wohnbevölkerung. In: Kecskes, Robert/Wagner, Michael/Wolf, Christian (eds.): Angewandte Soziologie. Wiesbaden: VS – Verlag für Sozialwissenschaften, pp. 21–45.

Blasius, Jörg (2008): 20 Jahre Gentrification-Forschung in Deutschland. In: Informationen zur Raumentwicklung 11/12, pp. 857–860.

Blasius, Jörg/Dangschat, Jens S. (eds.) (1990): Gentrification – Die Aufwertung innenstadtnaher Wohngebiete. Frankfurt am Main. Campus.

Blasius, Jörg/Friedrichs, Jürgen/Rühl, Heiko (2016): Gentrifikation in zwei Wohngebieten von Köln. In. Kölner Zeitschrift für Soziologie und Sozialpsychologie, 68, pp. 541–559.

Boterman, Willem/van Gent, Wouter (2023): Making the Middle-class City. The Politics of Gentrifying Amsterdam. New York: Palgrave MacMillan.

Bourdieu, Pierre (1986): The Forms of Capital. In: John G. Richardson (ed.), Handbook of Theory and Research for the Sociology of Education. New York. Greenwood Press, pp. 241–258.

Bourdieu, Pierre (1990): Structure, Habitus, Practice. In: Bourdieu, Pierre: The Logic of Practice. Cambridge: Polity Press, pp. 52-65.
Bourdieu, Pierre (1998): Ortseffekte. In: Göschel, Albrecht/Kirchberg, Volker (eds.): Kultur in der Stadt. Wiesbaden: VS Verlag für Sozialwissenschaften, pp. 17-25.
Brenner, Neil (1998): Between fixity and motion: accumulation, territorial organization and the historical geography of spatial scales. In: Environment and Planning D: Society and Space 16, pp. 459-481.
Buzek, Richard/Mießner, Michael (2022): Rent gap-getriebene kaskadenförmige Ausdehnung immobilienwirtschaftlicher Aufwertung entlang der Städte-Hierarchie? In: Glatter/Mießner (eds.) (2022a), pp. 107-125.
Christophers, Brett (2012): Anaemic Geographies of Financialisation. In: New Political Economy, 17, 3, pp. 271-291.
Clark, Eric (1987): The Rent Gap and Urban Change. Case Studies in Malmö 1860-1985. Lund: Lund University Press.
Clark, Eric (1991): Rent gaps and value gaps: Complementary or contradictory? In: van Weesep/Musterd (eds.) (1991), pp. 17-29.
Clay, Phillip L. (1979): Neighborhood Renewal: Middle Class Resettlement and Incumbent Upgrading in American Neighborhoods. Lexington: D.C. Heath.
Cresswell, Tim (2013): Geographic Thought. A Critical Introduction. Malden, MA/Oxford: John Wiley.
Dangschat, Jens S. (1988): Gentrification: Der Wandel innenstädtischer Wohnviertel. In Friedrichs, Jürgen (ed.): Soziologische Stadtforschung. Sonderheft 29 der Kölner Zeitschrift für Soziologie und Sozialpsychologie. pp. 272-292.
Dangschat, Jens S. (1990): Geld ist nicht (mehr) alles – Gentrification als räumliche Segregierung nach horizontalen Ungleichheiten. In: Blasius/Dangschat (eds.) (1990), pp. 69-91.
Dangschat, Jens S. (1991a): Gentrification – Indikator und Folge globaler ökonomischer Umgestaltungen, des sozialen Wandels, politischer Handlungen und von Verschiebungen auf dem Wohnungsmarkt in innenstadtnahen Wohngebieten. Hamburg: Universität Hamburg, unveröff. Habilitation.
Dangschat, Jens (1991b): Gentrification in Hamburg. In: van Weesep/Musterd (eds.) (1991), pp. 63-88.
Dangschat, Jens S. (2007): Raumkonzept zwischen struktureller Produktion und individueller Konstruktion. In: Ethnoscripts 9, 1, pp. 24-44.
Dangschat, Jens S. (2009): Symbolische Macht und Habitus des Ortes. Die „Architektur der Gesellschaft" aus Sicht der Theorie(n) sozialer Ungleichheit von Pierre Bourdieu. In: Fischer, Joachim/Delitz, Heike (eds.): Die Architektur der Gesellschaft. Theorien für die Architektursoziologie. Bielefeld: Transcript, pp. 311-341.
Dangschat, Jens S./Alisch, Monika (1995): Gentrification in Hamburg. Die ökonomische Aufwertung und kulturelle Umwertung dreier innenstadtnaher Wohngebiete. Hamburg: Universität Hamburg, unveröff. Forschungsbericht, gefördert durch Deutsche Forschungsgemeinschaft, Da 219/1-2.
Dangschat, Jens S./Blasius, Jörg (eds.) (1994): Lebensstile in den Städten. Opladen: Leske+Budrich.

Dangschat, Jens S./Friedrichs, Jürgen (1988): Gentrification in Hamburg. Eine empirische Untersuchung von drei Wohnvierteln. Hamburg: Gesellschaft für Sozialwissenschaftliche Stadtforschung.

Deleuze, Gilles/Guattari, Félix (1992): Tausend Plateaus – Kapitalismus und Schizophrenie II. Berlin: Merve.

Dolowitz, David P./Marsh, David (2000): Learning from abroad. The role of policy transfer in contemporary policy-making. In: Governance, 13, 1, pp. 5–23.

Eckardt, Frank (2018): Gentrifizierung: Forschung und Politik zu städtischen Verdrängungsprozessen. Wiesbaden: Springer VS.

Elam, J. Daniel (2019): Postcolonial Theory. In: Oxford Bibliographies.

Evans, Mark/Davies, Jonathan (1999): Understanding policy transfer. A multi-level, multidisciplinary perspective. In: Public Administration 77, 2, pp. 361–385.

Florida, Richard (2002): The Rise of the Creative Class. And How It's Transforming Work, Leisure and Everyday Life. New York: Verso.

Friedrichs, Jürgen (1996): Gentrification: Forschungsgegenstand und methodische Probleme. In: Friedrichs/Kecskes (eds.) (1996), pp. 13–40.

Friedrichs, Jürgen/Blasius, Jörg (2016): Gentrification in Köln. Soziale, ökonomische, funktionale und symbolische Aufwertungen. Opladen: Leske+Budrich.

Friedrichs, Jürgen/ Kecskes, Robert (eds.) (1996): Gentrification – Theorie und Forschungsergebnisse. Opladen: Leske + Budrich.

Ghertner, Asher (2015): Why Gentrification Theory Fails in 'Much of the World'?, City 19, pp. 552–563.

Glatter, Jan (2006): News from the Blind Men and the Elephant? – welche neuen Erkenntnisse bietet die jüngere Gentrificationforschung? In: Europa Regional 14, 4, pp. 156–166.

Glatter, Jan/Mießner, Michael (eds.) (2022a): Gentrifizierung und Verdrängung. Aktuelle theoretische, methodische und politische Herausforderungen. Bielefeld: Transcript.

Glatter, Jan/Mießner, Michael (2022b): Aktuelle Debatten in der deutschsprachigen Gentrifizierungsforschung – Zur Einleitung. In: Glatter/Mießner (eds.) (2022a), pp. 9–31.

Glatter, Jan/Mießner, Michael (2022c): Gentrifizierung und ihre Erforschung im deutschsprachigen Raum – Historische Entwicklungen. In: Glatter/Mießner (eds.) (2022a), pp. 33–54.

Haase, Annegret/Rink, Dieter (2015): Inner-city transformation between reurbanization and gentrification: Leipzig, eastern Germany. In: Geografie 120, 2, pp. 226–250.

Hackworh, Jason/Smith, Neil (2001): The changing state of gentrification. In: Tijdschrift voor economische en sociale geografie 92, 4, pp. 464–477.

Hamnett, Chris (1984): Gentrification and residential location theory: a review and assessment. In: Herbert, David T./Johnston, Ronald J. (eds.): Geography and the Urban Environment. London: Wiley, pp. 283–319.

Hamnett, Chris (1991): The Blind Man and the Elephant: An Explanation of Gentrification. In: van Weesep/Musterd (eds.) (1991), pp. 30–51.

Hamnett, Chris (2003): Gentrification and the middle class remaking of Inner London 1961–2001. In: Urban Studies 40, 12, pp. 2401–2426.

Hamnett, Chris/Randolph, Bill (1986): Landlord disinvestment and housing market transformation: the flat break-up market in London. In: Smith//Williams (eds.) (1986), pp. 121–152.

Helbrecht, Ilse (2016): Gentrifizierung in Berlin: Verdrängungsprozesse und Bleibestrategien. Bielefeld: transcript.

Holm, Andrej (2012): Gentrification. In: Eckardt, Frank (ed.): Handbuch der Stadtsoziologie. Wiesbaden: Springer SV, pp. 661–687.

Holm, Andrej (2014): Gentification. In: Belina, Bernd/Naumann, Matthias/Strüver, Anke (eds.): Handbuch kritische Stadtgeographie. Münster: Westfälisches Dampfboot, pp. 102–107.

Holm, Andrej (2020): Gentrifizierung in ostdeutschen Städten. In: Becker, Sören/ Naumann, Matthias (eds.): Regionalentwicklung in Ostdeutschland. Dynamiken, Perspektiven und der Beitrag der Humangeographie. Wiesbaden: Springer, pp. 309–320.

Huber, Florian J. (2013): Gentrifzierung in Wien, Chicago und Mexico Stadt. Qualitative Stadtforschung und internationale Vergleichbarkeit. In: Österreichische Zeitschrift für Soziologie, 38, pp. 237–257.

Janoschka, Michael (2022): Gentrifizierung, Finanzialisierung und Demokratie. Konzeptionelle Herausforderungen an kritische Stadtgeographien. In: Glatter/Mießner (eds.) (2022a), pp. 73–89.

Kadi, Justin (2019): Which Cities are Studied? Probing the Geographical Scope of 40 Years of Gentrification Research. In: Der öffentliche Sektor – The Public Sector 45, 1, pp. 48–54.

Kadi, Justin/Matznetter, Walter (2022): The long history of gentrification in Vienna, 1890–2020. In: City, online. https://doi.org/10.1080/13604813.2022.2054221.

Kadi, Justin/Verlic, Mara (eds.): Gentrifizierung in Wien. Perspektiven aus Wissenschaft, Politik und Praxis. Wien: Kammer für Arbeiter und Angestellte.

Läpple, Dieter (1991): Essay über den Raum: Für ein gesellschaftswissenschaftliches Raumkonzept. In: Häußermann, Hartmut/Ipsen, Detlef/Krämer-Badoni, Thomas/ Läpple, Dieter/Rodenstein, Marianne/Siebel, Walter (eds.), Stadt und Raum: Soziologische Analysen. Pfaffenweiler: Centaurus, pp. 157–207.

Lees, Loretta (2008): Gentrification and social mixing: towards an inclusive urban renaissance? In: Urban Studies, 45, 12, pp. 2449–2470.

Lees, Loretta/Slater, Tom/Wyly, Elvin (2008): Gentrification. New York: Routledge.

Lees, Loretta/Shin, Hyun B./López-Morales, Ernesto (2016): Planetary Gentrification. London: Polity.

Lefebvre, Henri (1968): Le Droit à la Ville.

Lefebvre, Henri (1991): The Social Production of Space. Oxford: Blackwell.

Ley, David (1980): Liberal Ideology and the Post-Industrial City. In: Annals of the Association of American Geographers 70, pp. 238–258.

Ley, Davis (1986): Alternative explanations for inner-city gentrification: A Canadian assessment. In: Annals of the Association of American Geographers, 76, pp. 521–535.

Ley, David (1996): The New Middle Class and the Remaking of the Central City. Oxford: Oxford University Press.

Maloutas, Thomas (2012): Contextual Diversity in Gentrification Research. In: Critical Sociology 38, 1, pp. 33–48.

Marcuse, Peter (1986): Abandonment, Gentrification and Displacement: The Linkages in New York City. In: Smith/Williams (eds.), pp. 153–177.

Massey, Doreen (1993): Power-geometry and a progressive sense of place. In: Bird, Jon/Curtis, Barry/Putnam, Tim/Robertson, George/Tickner, Lisa (eds.): Mapping the futures. London: Routledge, pp. 59–69.

Mayer, Margit (2012): The "right to the city" in urban social movements. In: Brenner, Neil/Marcuse, Peter/Mayer, Margit (eds.): Cities for People not for Profit – Critical Urban Theory and the Right to the City. New York: Routledge, pp. 63–85.

Mayntz, Renate/Scharpf, Fritz W. (1995): Der Ansatz des akteurzentrierten Institutionalismus. In: Mayntz, Renate/Scharpf, Fritz W. (eds.): Gesellschaftliche Selbstregelung und politische Steuerung. Frankfurt am Main: Campus, pp. 39–72.

McCann, Eugene/Ward, Kevin (2012): Policy Assemblages, Mobilities and Mutations. Towards a Multi-Disciplinary Conversation. In: Political Studies 10, 3, pp. 325–332.

Phillips, Martin (1993). Rural Gentrification and the Processes of Class Colonisalisation. In: Journal of Rural Studies 9, 2, pp. 123–140.

Reick, Philipp (2018): Gentrification 1.0: Urban Transformation in Late-19th-Century Berlin. Urban Studies 55, 11, pp. 2542–2558.

Robertson, Roland (1995): Glocalization: Time-Space and Homogeneity-Heterogeneity. In: Featherstone, Mike/Lash, Scott/Robertson, Roland (eds.): Global Modernities. London: Sage Publications, pp. 25–44.

Rose, Damaris (1984): Rethinking gentrification: beyond the uneven development of Marxist urban theory. In: Society and Space, 2, pp. 47–74.

Sassen, Saskia (1991): The Global City: New York, London, Tokyo. New Jersey: Princeton University Press.

Slater, Tom (2017): Planetary Rent Gaps. In: Antipode 49, 1, pp. 114–137.

Smith, Neil (1979): Toward a theory of gentrification: a back to the city movement by capital, not people. In: Journal of the American Planning Association, 45, pp. 538–548.

Smith, Neil (1984): Uneven Development. Nature, Capital and the Production of Space. Cambridge/Mass: Basil Blackwell.

Smith, Neil (1986): Gentrification, the frontier, and the restructuring of urban space. In: Smith/Williams (eds.) (1986), pp. 15–34.

Smith, Neil (1987): Gentrification and the rent gap. In: Annals of the Association of American Geographers, 77, pp. 462–478.

Smith, Neil (1991): On gaps in our knowledge of gentrification. In: van Weesep/Musterd (eds.) (1991), pp. 52–62.

Smith, Neil (1996): The New Urban Frontier. Gentrification and the Revanchist City. London/New York: Routledge.

Smith, Neil (2002): New Globalism New Urbanism: Gentrification as Global Urban Strategy. In: Antipode 34, pp.427–450.

Smith, Neil/Williams, Peter (eds.) (1986): Gentrification of the City. London: Routledge.

Stockdale, Aileen (2010): The Diverse Geographies of Rural Gentrification in Scotland. In: Journal of Rural Studies 26, 1, pp. 31–40.

Tuan, Yi-Fu (1977): Space and Place: The Perspective of Experience. Minneapolis, MN: University of Minnesota Press.

Üblacker, Jan (2018): Gentrifizierungsforschung in Deutschland: eine systematische Forschungssynthese der empirischen Befunde zur Aufwertung von Wohngebieten. Berlin/Opladen: Budrich Uni Press.
van Weesep, Jan/Musterd, Sako (eds.) (1991): Urban Housing for the Better-Off: Gentrification in Europa. Utrecht: Stedelijke Netwerken.
Wyly, Elvyn (2019): The Evolving State of Gentrification. In: Tijdschrift voor Economische en Sociale Geografie 110, 1, pp. 12–25.
Zukin, Sharon (1982): Loft Living. Culture and Capital in Urban Change. New Brunswick: Rutgers University Press.
Zukin, Sharon (1987): Gentrification: culture and capital in the urban core. In: Annual Review of Sociology, 13, pp. 129–147.

CARME – The Rise and Fall of a Sociological Space in the East

Karl M. van Meter[1]

Abstract

When Mikhail Gorbachev in 1985 came to power in the USSR, his wife, Raisa, often described as an activist and a philanthropist interested in fostering new talent, decided she wanted to defend a doctoral thesis in sociology, which was then a sub-branch of Marxist-Leninist Philosophy at the Soviet Academy of Science. Mikhail Gorbachev had an independent Institute of Sociology created in May 1988 (formerly the Institute for Concrete Social Research) under the direction of Vladimir Jadov who became the director of Raisa's thesis. To establish the bona-fide scientific nature of sociology in the USSR, it was necessary to organize an internationally recognized sociology conference in the USSR, but the German Democratic Republic (GDR) had already a lead in this endeavor through German relations with the International Sociological Association (ISA) Research Committee 33 (RC33), "Logic and Methodology". The USSR seems to have "pulled rank" on the GDR and with UNESCO so that RC33 chaperoned the first – and the only – international sociology conference in the Soviet Union, Moscow (24–27 October 1988), before the first – and the only – international sociology conference in the GDR (Holzhau, 2–6 October 1989). CARME work was used as the "litmus test" in both cases to show the world that empirical sociology existed in the East, but unfortunately not for long [the Berlin Wall fell on 9 November 1989, Germany was reunified on 3 October 1990, the Soviet Union was dissolved on 26 December 1991].

1. Sociology in the Soviet Union

Well before Mikhail Gorbachev came to power in the USSR in 1985, Soviet and other Eastern European sociologists were doing internationally recognized work and exchanging with their Western colleagues. Artur Meier at Humboldt University in East Berlin was a recognized international specialist in the sociology of education and an International Sociological Association (ISA) vice president in the late 1980s. Rudolf Andorka in Hungary was working on social

---

1 The author was one of the founding editors of the *Bulletin of Sociological Methodology/ Bulletin de Méthodologie Sociologique (BMS)* in 1983, and during the period of the events presented in this article, he was Secretary and then President of RC33 of the ISA, as well as ISA Executive Committee member for two tenures

mobility and in 1984 became chair of sociology at the Budapest University of Economic Science. Horst Berger helped found the Soziologisch-Methodische Zentrum (SMZ) and in 1988 became deputy director of the Institute of Sociology and Social Policy (ISS) of the East German Academy of Sciences. Soviet sociologist Vladimir G. Andreyenkov was a vice president of ISA's RC33 in the late 1980s and he published several articles in the the RC33-associated *Bulletin of Sociological Methodology/Bulletin de Méthodologie Sociologique (BMS)*, including "Analysis and Questionnaire of the Survey 'Public Opinion in the Soviet Union and West Germany'":

> Following the previous joint Soviet-American and Soviet-French public opinion polls, the author and the Institute of Sociology of the Academy of Sciences of the USSR organized the first Soviet-West German public opinion poll during the October 1988 visit of Helmut Kohl to the Soviet Union. The questionnaire is presented along with a brief analysis of the results. (Andreyenkov, 1989)

The *BMS* also presented Andreyenkov's "Cross-national American-Soviet Teenager Survey: The Future and Nuclear War" (Andreyenkov, 1987), consisting largely of statistical cross classifications, and one of Andreyenkov's short books based on international cooperation: Vladimir G. Andreyenkov and Valeriy A. Mansurov, *The World as Seen by Contemporary People* (1988), subtitled "Preliminary Results of the Study of the Soviet and French Citizens' Relation to Home and International Policy Problems":

> In October 1987, the Institute for Sociological Research of the USSR Academy of Sciences and IPSOS (France) carried out a joint survey sponsored by the USSR Committee "Gosteleradio" and the first channel of the French Television TF1. (Andreyenkov and Mansurov, 1988, in *BMS*, n. 19, July 1988: 9)

From the beginning, RC33 had had Soviet and later Russian members who were interested in sociological methodology. I had learned through colleagues that the Soviet/Russian weekly *Argumenti y Fakti*, in Moscow, was actually publishing questionnaire survey results concerning health care, housing and other typical social issues. The problem was that "sociology" didn't officially exist since such work was part of Marxist-Leninist Philosophy at the Academy of Science. The crypto-sociologists at the Academy of Science were actually constructing typical pencil-and-paper survey questionnaires, which they would pass on to *Argumenti y Fakti* that would publish them as "tear-out" pages of the weekly for eventually interested respondents who could fill out the questionnaire and send it back to the journal by postal mail, which we now call "snail mail". Once the survey terminated, the finished questionnaires were then given by *Argumenti y Fakti* to the crypto-sociologists for data analysis, since the latter had access to the few personal computers that existed and the necessary software, such as pirated versions of SAS that Eastern European technicians had managed to "crack" the encrypted key. The results were discussed between the crypto-sociologists and *Argumenti y Fakti* journalists who would then write up and publish the results in the weekly. The weekly as also known at the time as the only Soviet journal with a credible specialist on the KGB who would often publish critical evaluations of the KGB's work.

2. International contacts

Already in 1987, RC33 President Manfred Küchler, formerly of the sociological ZUMA research center in Mannheim, was in contact with East German colleagues Artur Meier and Horst Berger:

> The [RC33] membership renewal process is completed, but some problems remain. Most notably, there are payment problems for colleagues residing in countries with no freely convertible currency. VP Andreyenkov has informed me that quite a number of colleagues are interested in joining, but are experiencing payment problems.
>
> I did a fair amount of travel for RC33 business during the first months of 1987 including . . . a visit to the Institute of Sociology [and Social Policy] at the Humboldt University in Berlin, East Germany, in late March . . . In Berlin I enjoyed very stimulating discussions with students and fellow scholars on a broad range of methodological issues. Chances are very good that our East German colleagues will host an RC33 conference (with special focus on the use of microcomputers in social research as originally proposed by Horst Berger). Matter of fact, I just received a letter by ISA Vice President Artur Meier confirming these plans. The final date has not been set: it will either [be] in the fall of 1988 or in 1989. And I would like to specially welcome Manfred Lindtner and Joachim Rudolph of the Institute of Sociology [and Social Policy] as new RC33 members. (Küchler, 1987)

In terms of technological progress, the publication of the above report represented something that was then considered "nearly impossible":

> As mentioned in the "Editorial" of this issue, the above "President's Report" by Manfred Küchler was keyed in on a terminal at Florida State University, transmitted free of charge by electronic mail to our local node (FRORS31) in care of our userid (SDC200), printed out on a high quality laser printer, and used directly for the layout of this *RC33 Newsletter*. As

far as any of us know, this seems to represent a technological "première" for scholarly communication. (van Meter, 1987) [at the time, the Internet did not exist and the text was transmitted via Bitnet in the US to EARN in Europe]

And there were also "professors of sociology" at the Soviet Academy of Sciences inside the branch of Marxist-Leninist Philosophy, one of whom was Vladimir A. Jadov who later was on the Executive Committee of the ISA with me in 1990-1994. He was a specialist in the sociology of labor and also economic sociology, plus author of Russia's first textbook in sociological methodology – *Strategy of Sociological Research* – which is still of fundamental importance in Russian sociology curriculum. Vladimir, a short but wide-shouldered amiable man with a bright smile on a finely-featured smallish head, often asked me if there was any miracle medicine in the West for his ailing shoulder that kept him from using his axe to chop firewood, which was the only means of heating on his farm in Simuna, Estonia. I didn't know any such miracle medicine for him, but he did have something of great interest for me, and for RC33 and for the *BMS*: one of his former students, Raisa Maximovna Gorbacheva, the rather active wife of whom would be the last Soviet leader, Mikhail Gorbachev, had wanted to defend a "real" sociology doctorate. Raisa did defend her sociology thesis under Jadov at the Institute for Sociological Research and it was apparently decided that to establish the bona-fide scientific stature of sociology in the USSR, it was necessary to organize an internationally recognized sociology conference that would fit in very well with ongoing processes in the USSR.

3. A process of transformation and the planning of two conferences

In May 1988, the Soviet Central Committee decided to transform the Institute for Concrete Social Research, then under the authority of Marxist-Leninist Philosophy at the Soviet Academy of Science, into an independent Institute for Sociological Research:

> Among the first practical decisions resulting from this political decision, was to change the name of the Institute for Concrete Sociological Research to Institute of Sociology and name Vladimir A. Jadov temporary director in charge of reorganizing the Institute. Prof. Jadov is Vice-President of the Soviet Association of Sociology. The Institute has been given its own budget and Prof. Jadov has the right, and the institutional autonomy, to use the budget as he sees fit. Therefore, the Institute no longer has to refer to higher scientific or political instances in order to make its policy decisions. The reorganisation of the Institute, which has been under

way since the summer 1988, will be presented soon to higher authorities [...] These developments in Moscow have their equal in Kiev and in Novosibirsk where apparently new institutes of sociology and scientific sociological reviews are to be established soon. Indeed, a new and reinvigorated sociology has an important role to play in the development of Mikhail Gorbachev's policy of "glasnost". (van Meter, 1989)

That "reorganisation of the Institute, which has been under way since the summer 1988" had already been felt at RC33:

> Secondly, Vice President Vladimir Andreyenkov has advised me that the Soviet Sociological Association, as well as the Academy of Sciences, have supported his suggestion to hold an RC33 symposium sometime in October or November of 1988 [...] Thirdly, a workshop on "Computer-aided methods in social research" [CASOR'89] will be held in the German Democratic Republic (GDR) in the fall of 1989. This workshop was proposed by Professor Horst Berger and has found the support of the Academy of Sciences to the GDR [...] Vice President Vladimir Andreyenkov and Professor Horst Berger will be in close touch to coordinate these two meetings. (Küchler, 1988)

So, suddenly in January 1988, the USSR had become interested in organizing an international sociology conference before the GDR and to have it take place just a few months later, but the first real "Call for Papers" the *BMS* received was for the GDR CASOR'89 conference:

> Appels/Calls – Computer Aided Sociological Research – CASOR'89 – In connection with the Research Committee on Logic and Methodology (RC33 of the International Sociological Association), the Institute of Sociology and Social Politics of the Academy of Sciences of the GDR will organize a workshop on new methods or experiences in computer aided sociological research. The conference will be held in autumn 1989 over a period of five days, probably in Holzhau (Berger, 1988) [*BMS*, n. 18, April 1988: 59]

The Soviet "Call for Papers" came later and was published in the following issue of the *BMS*, giving interested parties only two months to prepare for an international conference:

- International Symposium on Methodological Aspects of Empirical Research in Sociology

- October 24–27, 1988, Moscow, USSR

- Presentation – Carrying on the tradition of interdisciplinary conferences in Amsterdam (1984) and Yugoslavia (May 1988) and at the suggestion of RC33 of the International Sociological Association, the Institute for Sociological Research and the Soviet Sociological Association of the USSR Academy of Sciences are planning to organize an International Symposium on Methodological Aspects of Empirical Research in Sociology . . . (..., *BMS*, n. 19, July 1988: 15–16)

Essentially the Soviet Union had jumped on the bandwagon, displacing East Germany and pushing back the latter's international conference by a full year. At the time, East Germany did not have the equivalent of the USSR's *Argumenti y Fakti* where actual modern survey research was being quietly carried out and where there was actual criticism of the local intelligence service; that is to say the KGB. Its East German equivalent, the Stasi, and the GDR Central Committee had the country so tightly under control that an *Argumenti y Fakti* was then impossible. Nonetheless, there was enough interest and momentum in East German academic *milieux* to have first taken the initiative to organize an international sociology conference.

Manfred Küchler, as RC33 President, and I, as RC33 Secretary and *BMS* Editor, decided that it was more important to make sure that the Soviet conference took place even if it was clearly "a rushed job", but that both meetings required respecting UNESCO and ISA requirements for an open international conference. As we mentioned in the opening abstract:

> The USSR seems to have "pulled rank" on the GDR and with UNESCO so that RC33 chaperoned the first – and the only – international sociology conference in the Soviet Union (Moscow, 24–27 October 1988), before the first – and the only – international sociology conference in the GDR (Holzhau, 2–6 October 1989).

4. Analyzing the Soviet "power elite"

While both the Soviet and the East German conferences were being organized, there had been an ongoing discussion in the transatlantic social network analysis community concerning the role of ideology, particularly political ideology, in sociology methodology. Were "Western" methods, that were currently being used in Western sociological research with Western data sources, also applicable in Eastern European sociological research? One of the cases cited in this debate in the journal *Connections* was the social network analysis of biographies of leading American political and economic personalities which supposedly formed what C. Wright Mills called "The Power Elite". With French colleagues, we maintained that similar work in France – showing that the famous "One Hundred Families" ran everything in France – had been for many years a French Communist Party political recruitment theme and it would be dangerous to use similar methods in any credible current French sociological research (van Meter, 1987).

Since I had worked with Michel Tatu, the daily Le Monde's specialist on Eastern Europe, I had access to his vast *Sovt* data base which included detailed biographies of all Soviet leaders. Tatu had compiled all editions of the *Great Soviet Encyclopedia* which, from 1957 to 1990, was published annually in the form of the *Yearbook of the Great Soviet Encyclopedia* with up-to-date articles about the Soviet Union, Soviet leaders and all countries of the world. These detailed biographies of Soviet leaders had a marked tendency to change from year to year as the individual leader's "political capital" changed over time, and some time even disappeared entirely. A similar evolution of individual leaders' "political capital" had been noted in the People's Republic of China. In Tatu's well-structured data base, all entries included published references, but I once asked where did the very few non-referenced entries come from. There was no clear answer, but I was made to understand that perhaps the CIA had some information that Michel Tatu didn't have.

With access to Tatu's data base, our group of French social network colleagues decided to use "Western" methods to analyze it, and of course the methods included factorial correspondence analysis and hierarchical ascending classification analysis. After initial explorations, we decided that the best approach would be to use the geographical key words in the official biographies of all members of the Central Committee of the Communist Party of the USSR during the mandate of Mikhail Gorbachev, trying to see if we could find "cliques", or a "power elite" like in the US social data.

This work was underway when the Moscow conference was being organized. What a wonderful occasion to "test" the international acceptability of the Moscow "sociology" conference by submitting our work on the possible structure of the Central Committee for presentation! Our work and the methods could act as a "litmus test" to prove to UNESCO and other international organizations that the Moscow conference was indeed an "international sociology conference" in its own right.

This was done, the presentation was accepted, and the conference took place on 24–26 October 1988 in Moscow: the "International Symposium on Methodological Aspects of Empirical Research in Sociology", the first and the only international sociology conference to have ever taken place in the USSR, which disappeared three years later on 26 December 1991. There were some 30 participants from Western Europe and North America and approximately 50 came from the Soviet Union and other Eastern European countries (..., *BMS*, n. 21, January 1989: 8–23). After an opening Soviet presentation, it was with some apprehension on 24 October 1988, and in front of that audience, that I presented "East Meets West – Official Biographies of Members of the Central Committee of the Communist Party of the Soviet Union between 1981 and 1987, analyzed with Western Social Network Analysis Methods" (van Meter et al., 1989). When I finished the presentation, there was a silence in the vast, cold conference hall until a Soviet participant raised his hand and simply asked: "OK, and now what happens?"

I was not at all prepared for such easy acceptance, but the answer to "now what happens" seemed rather obvious from the two-dimensional factorial correspondence graphic that we had generated. There were essentially two large but unequal spaces or clusters of geographical place names. The smaller space was situated along the left-hand side of the first axis and consisted largely of geographical names associated with Crimea and the Ukraine and the older colleagues of Leonid Brezhnev. The larger space covered the entire right-hand side of the graphic, extending from the bottom with Latvia and other names associated the Baltic Republics, all the way up to the top where the names were associated with the Central Asian Republics (Uzbekistan and Azerbaijan). In the dense middle part, one could find names and regions associated with Moscow or with Leningrad (later St. Petersburg). Perpendicular to this two-dimensional graphic, there was a third dimension largely determined by the geographical name "Stavropol". We were obliged to ask Eastern European specialists at Sciences Po in Paris what was the significance of Stavropol. They just laughed at us. Stavropol was Mikhail Gorbachev's home town and seat of power situated between the extremes, just like Moscow and Leningrad, and obviously situating itself as a new pole of power independent of Moscow and Leningrad.

Figure 1: Factoral analysis of geographical names (axes 1 and 2)

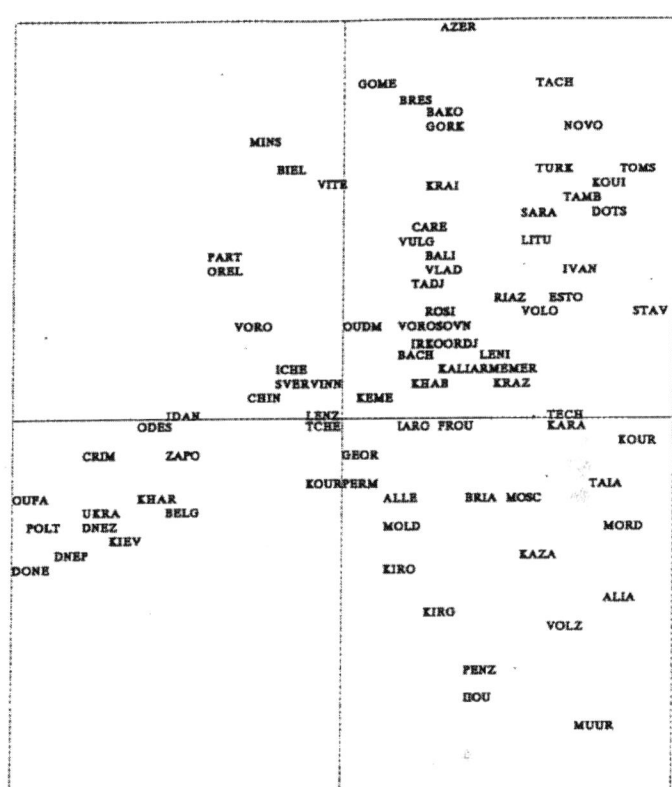

(van Meter et al., 1989: 37 – One should note that Stavropol (STAV), which largely constitutes the third vertical axis, appears in this two-dimensional graphic on the far right-handed border just above the first axis.)

So the answer to the question "now what happens" was obviously "try to hold it all together" and that was going to be very difficult. Another Soviet participant confirmed that view and suggested an example of the difference between the rather well-run and more modern Baltic Republics and, for example, Uzbekistan. When I asked for specific details, I was told that under Gorbachev, international economic opportunities had been encouraged and Uzbekistan had been allowed to put an entire recent harvest of its high-quality cotton on the international market. There had been several buyers and ownership had gone from one big buyer to another. At one point, an American businessman offered to purchase the entire harvest, but he wanted to visit Uzbekistan first and inspect the cotton. He flew to Uzbekistan and found that since it had been harvested the cotton had been stored outside, and often under the rain, and was completely rotten and worthless. Senior authorities in Stavropol, or Moscow, or Leningrad, had put the entire Uzbek Central Committee before a firing squad for corruption and wasting an entire cotton crop of that Republic.

5. Experiencing the first and last international sociological conference in the GDR

As for planning and organization, the Moscow sociology conference was only slightly better than the Uzbek Central Committee, and there was never a published volume of the presentations. Our article with its graphic and under the same title as the presentation was later published in *Connections* (van Meter et al., 1989). We chose *Connections* for publication since we authors had decided that it was more important to publish it in the US social network analysis community than to keep the article for the more international *BMS* readership. However, the *BMS* did publish a three-page review of the Moscow conference:

> Moscow, 24–26 October 1988 – Galina N. Saganenko (Inst. Socio-Economic Problems, Leningrad), "On the Problem of Analyses of Interdependence of Social Factors". Karl M. van Meter, "East meets West" [...] there was a total of 31 presentations, 30 Western and 50 Eastern participants [with the full list of presentations; ..., *BMS*, n. 21, January 1989: 21–23]

This was the only report made of that conference and there was never any official publication that we know of. But this offered East German sociology the possibility to "correct the situation" and "do better", and at the same time, be able to "get even" with the USSR for taking away from them the privilege of organizing the first international sociology conference in Eastern Europe. However, the GDR had its own problems organizing such an international meeting in Holzhau in October 1989.

At that time, West German colleague Jost Reinecke had recently been to the East Berlin celebration for East German's Republic Day on 7 October 1989, the 40th anniversary of the founding of the GDR, and there were serious street demonstrations. Jost learned how to take pictures of the action and still avoid the Stasi by ducking into cafés. Throughout Eastern Europe there had been protests during the preceding weeks, and Hungary and Czechoslovakia decided to allow East Germans to travel freely across their borders. Once in Hungary, East Germans could walk into Austria. Holzhau was very near the Czech border. In Czechoslovakia, fleeing East Germans were able to take refuge in the West German embassy in Prague, from there some of them were transported in closed and locked trains through East German to the West. While this was happening, in Holzhau, Jost Reinecke was sharing a room with Jörg Blasius who was to present one of his first factorial correspondence analysis papers – "Correspondence Analysis – Theory and Application" (Gladitz and Troitzsch, 1990: 65–74) – with its "description of this French method and its possible extension to a wide range of sociological studies".

I was sharing a room with Peter Mohler from the Mannheim ZUMA research center. Largely unaware of what was happening on the international scene, one morning I got up early and went running uphill into the forests and soon found I had crossed a small stream and was in Czechoslovakia. That didn't bother me, but the threatening wooden guard towers I came across did indeed bother me. Continuing running at a slower pace, I realized the towers were crooked and collapsing, having been abandoned long ago. Happy with that discovery, I ran along, jumping back and forth between Czechoslovakia and East German many times, and, on my way back down, greeting a group of serious young women running uphill. At breakfast, an elder lady "organizer" rushed to my table with Jost, Jörg and Peter and, looking very serious, asked me: "Were you running in the forest this morning?" I answered, "Yes". "Did you at least have your passport with you?" "Of course!" "Well then, everything is fine." I, of course, didn't have any documents on me while jogging. And that was not the end of your "misbehavior" at Holzhau.

The conference organizers thought it would be a good idea to have all of us visit nearby Freiberg and its Technische Universität Bergakademie which is considered "the world's oldest university for mining science." Of particular interest was the work of Friedrich Mohs who invented the Mohs scale of mineral hardness based on the difficulty of scratching a mineral's surface. First created in 1812, the scale has 10 levels of hardness based on whether or not different materials or minerals can scratch one or the other. It is considered the only scientific measurement system that does not involve numerical measurement since it is only a totally order set.

After that visit and a brief view of the well-maintained central plaza with its typical German buildings, conference participants were put back into a bus to return to Holzhau when Jost Reinecke suggested to Jörg Blasius and to me that we stay behind and very discreetly visit the rest of the town. The first thing Jost pointed out was that the central plaza that was often shown to foreign visitors was about the only well-maintained part of the town. Only a few dozen meters down one of the streets leading on to the central plaza, you could see the ruins of a building that had collapsed some time ago and its debris still covered the sidewalk and part of the street. Foreign visitors were not permitted to look down that street or any others. You were only allowed to appreciate the impeccable city center. Foreign dignitaries were usually driven into the city center in cars with smoked glass windows and low ceilings so they could only look out at street level and not see anything above or down side streets. Along that route to the central plaza for visiting dignitaries, the ground-level shops and houses were in fairly good condition. Elsewhere, that was not the case.

On one street, Jost pointed out a traffic cone marking a spot in the road that was to be repaired. The cone had been there for several years and, because the repairs hadn't been done, the building just next to it had been condemned and closed: it was Freiberg's only independent theater. Jost led us to the front door of an ordinary house, knocked on the door and whispered something to the person who had just slightly opened the door. We three were admitted, entered an empty hallway and then entered the first room to the left. In a very subdued atmosphere, small groups of young adults were sitting around small individual tables, drinking coffee or beer, and talking together. It was a clandestine café and Jost said there were several places like that in the city.

We continued "sneaking around town" and visiting its unofficial sights until late at night. Finally, we decided that we should get back to Holzhau before our absence became known to the "organizers" and cause a diplomatic incident. We went to the central station and found there was no form of public transportation available at night and there was a long line of people waiting at the taxi stand. We carefully snuck back into the taxi parking lot where taxi drivers were snoozing in their cars. We found a driver who seemed to be awake and started talking with him. I don't remember the exact details, but he understood our situation and we understood his: with a fixed salary, why go out of your way to take stranded shoppers back home at night? With a bit of financial encouragement, and I believe some foreign currency, he agreed to drive us back to Holzhau.

We three piled into the taxi and off we went. At one point, we mentioned that he was driving rather fast down the highway through the forests. He laughed and told us not to worry. "There aren't any police for miles around because they are all out chasing refugees through the forest and over the borders." He was right, and we got back to Holzhau without seeing a single police officer. The whole atmosphere in Freiberg and on the road was one of "fin de régime" or in American English, "it's all over now", and the Berlin Wall "opened" or "fell" on 9 November 1989, just one month and three days later.

6. "East Meets West" and what became of it

Once back in Paris, the *BMS* got down to writing up a report on the Holzhau meeting. The CASOR'89 conference report was published in the *BMS*, n. 25, December 1989: 13-16. It included a full list of the 31 presentations, including the "Theory and Application of Correspondence Analysis", by Jörg Blasius, and "French Sociologists Analyzed by French Sociological Methods", by colleagues and me – "a multimethod analysis using classification methods and factorial correspondence methods on data about French sociologists". However, the report on Holzhau included neither our "East Meets West" article, nor Joachim Rudolph's "Reanalyse eines Datensatzes von Karl M. van Meter" (Rudolph, 1990) based on the "East Meets West" article and the reanalysis of our data. Note clearly that neither of these two last articles had been presented at Holzhau.

Less than a year later, Akademie-Verlag in Berlin published the 526-page CASOR'89 book *Computer Aided Sociological Research* (Gladitz and Troitzsch, 1990) with the indication "Draft as of 2nd July 1990". The book included a detailed report of the conference in English and in German, the titles of the presentations, the names of the authors, and the full presentations. The book also included a copy of our original "East Meets West" article (van Meter et al., 1990), followed by Rudolph's article, both of which were not on the Holzhau program and not presented in Holzhau. The book was reviewed in the *BMS* (n. 28, September 1990: 15–17). I was told that as a compliment to me and my work, the book included the "East Meets West" article from the Moscow conference and that I was the only author with two articles in the book. I said thank you but did not check it out until I stared writing the present article. The text of the article was exactly the same as my original 1989 *Connections* "East Meets West", but I noticed that the inter-titles of the book's copy of the original article were slightly modified by the publisher, probably for format continuity, but the "Factoral analysis of geographical names" (Figure 1) had been modified to become the book's graphic (Figure 2) was substantially different from our original graphic above and must have been based on an East German reanalysis of our data, perhaps similar to that for the following article in the book – Joachim Rudolph's "Reanalyse eines Datensatzes von Karl M. van Meter".

Below is the book's graphic that accompanied that version of "East Meets West", and you can easily note the difference with the original graphic above and the greatly increased factorial contributions for Stavropol, Latvia and Riga, which is the capital and largest city of Latvia. This looks strangely like a graphic translation of my verbal presentation in Moscow and the audience's consensus that Stavropol, with its role as Gorbachev's center of power, and Latvia and the other Baltic Republics, with their "Westernized" economies, represented a significant change in contrast to the slow-moving, older Soviet "Establishment" that was far from enthusiastic in backing Gorbachev's policies of *perestroika* ("restructuring" in Russian) and *glasnost* ("openness"). Now with more than thirty years of history behind us, this certainly looks like wishful thinking concerning the future of the Soviet Union~Russia. The August 1991 putsch in Moscow deposed Gorbachev and ended his of *perestroika* and *glasnost* policies, Germany was reunified in October 1990, the Soviet Union was dissolved in December 1991. Latvia, along with Lithuania, Bulgaria, Estonia, Romania, Slovakia and Slovenia joined NATO in 2004, and last year, Putin's Russia invaded Ukraine. And in this strange new world, the original "East Meets West" graphic has strangely disappeared from the *Connections* Web site along with the entire volume 12, number 3 issue, luckily saved by Dutch researchers (see van Meter et al., 1989 below).

Figure 2: Factor analysis of geographical names (axes 1 and 2)

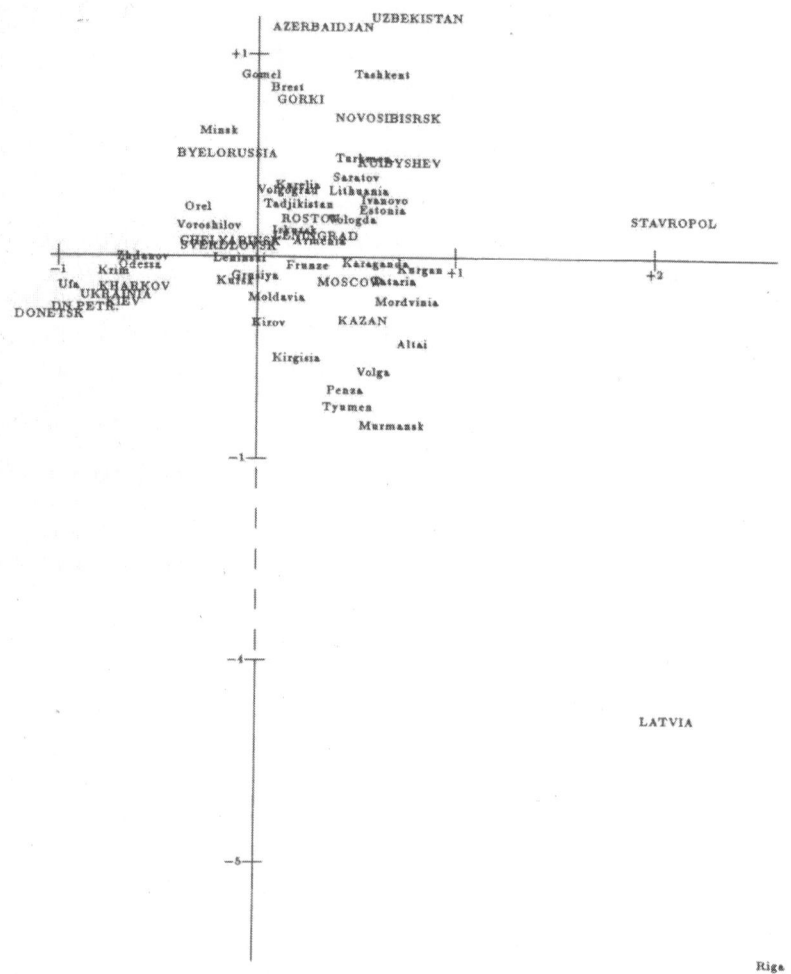

(van Meter et al., 1990, page 230)

References

...[unsigned text] (1988). In Bulletin of Sociological Methodology/Bulletin de Méthodologie Sociologique, 19, pp. 15–16.

...[unsigned text] (1989). In Bulletin of Sociological Methodology/Bulletin de Méthodologie Sociologique, 21, pp. 21–23.

...[unsigned text] (1989). In Bulletin of Sociological Methodology/Bulletin de Méthodologie Sociologique, 21, pp. 8–23).

Andreyenkov, Vladimir G. (1987): Cross-national American-Soviet Teenager Survey: The Future and Nuclear War. In: Bulletin of Sociological Methodology/Bulletin de Méthodologie Sociologique, 15, pp. 14–16.

Andreyenkov, Vladimir G. (1989): Analysis and Questionnaire of the Survey "Public Opinion in the Soviet Union and West Germany". In: Bulletin of Sociological Methodology/ Bulletin de Méthodologie Sociologique, 22, pp. 4–10.

Andreyenkov, Vladimir G./Mansurov, Valeriy A. (1988): The World as Seen by Contemporary People – Preliminary Results of the Study of the Soviet and French Citizens' Relation to Home and International Policy Problems. Moscow: Institute for Sociological Research, Academy of Sciences of the USSR, Moscow, 41 pp.

Berger, Horst (1988). In: Bulletin of Sociological Methodology/Bulletin de Méthodologie Sociologique, 18, p. 59.

Gladitz, Johannes/Troitzsch, Klaus G. (1990): Computer Aided Sociological Research. Berlin: Akademie-Verlag.

Küchler, Manfred (1987): President's Report. In: RC33 Newsletter, Bulletin of Sociological Methodology/Bulletin de Méthodologie Sociologique, 15, page 6.

Küchler, Manfred (1988): President's Report. In: RC33 Newsletter, Bulletin of Sociological Methodology/Bulletin de Méthodologie Sociologique, 17, pp. 7–8.

van Meter, Karl (1987a): Secretary's Report. In: RC33 Newsletter, Bulletin of Sociological Methodology/Bulletin de Méthodologie Sociologique, 17, p. 9.

van Meter, Karl M. (1987b): Ideology and Methodology: Network Analysis in the United States and France. Connections, 10/2, pp. 106–109.

van Meter, Karl M. (1989): Secretary's Report – The Moscow Meeting and Recent Developments in Soviet Sociology. Bulletin of Sociological Methodology/Bulletin de Méthodologie Sociologique, 21, pp. 8–9.

van Meter, Karl M./Cibois, Philippe/Mounier, Lise/Jenny, Jacques (1989): East Meets West – Official Biographies of Members of the Central Committee of the Communist Party of the Soviet Union between 1981 and 1987 Analyzed with Western Social Network Analysis Methods. In: Connections, 12/3, pp. 32–38, available at https://www.researchgate.net/profile/Meindert-Fennema/publication/241883517_Dutch_policy_networks_in_the_decolonization_of_Indonesia/links/54735e610cf216f8cfaff067/Dutch-policy-networks-in-the-decolonization-of-Indonesia.pdf#page=32 – the original copy on the Connections Web site has disappeared.

van Meter, Karl M./Cibois, Philippe/Mounier, Lise/Jenny, Jacques (1990): East Meets West – Official Biographies of Members of the Central Committee of the Communist Party of the Soviet Union between 1981 and 1987 Analyzed with Western Social Network Analysis Methods. In: Gladitz, Johannes/Troitzsch, Klaus G (eds.): Computer Aided Sociological Research. Berlin: Akademie-Verlag, pp. 223–232.

Social Milieus in Urban Space – A Comparison of Berlin, Singapore and Nairobi

Nina Baur and Elmar Kulke

Abstract

Relating to German sociologist Jörg Blasius' work, the authors investigate lifestyles (more specifically eating and housing practices of urban middle-class residents) and ask how lifestyles and urban structures are related in different social contexts. By comparing three cities of comparable sizes – Berlin (Germany), Singapore and Nairobi (Kenya), the paper illustrates how the middle classes ideal-typically relate to other social classes, how they position themselves in urban space, how they do distinction by eating and housing practices, and what effects this has on the organization of the food market.

1. Lifestyles, Housing and Urban Space as Markers of Social Inequality

How social inequality is (re-)produced, how it is expressed and what consequences it has, have been key questions sociology has been addressing since the discipline's beginning. Several key dimensions of social inequality exist – amongst them class (socio-economic status), race and ethnicity, gender, and age – which often intersect. It is an empirical question, in which social sphere and specific historical and spatial context which of these dimensions is most dominant in reproducing social inequality. Classical social inequality research has shown that in Europe and North America, in most empirical fields, social class is the dominant dimension of inequality and intersects with the other dimensions. Social inequality research also has demonstrated that the key resources for reproducing class inequality are property and income from paid work, the key indicators for social class being income, occupational status and education, which are empirically strongly linked up to today (Müller-Schneider 1996; Groh-Samberg et al. 2020).

The key argument of sociological lifestyle researchers such as Thorstein Veblen (1899), Pierre Bourdieu ([1979] 1984), Norbert Elias ([1939] 2000) or Gerhard Schulze (1992) is that in everyday life, you do not see how much people own or earn – therefore, in order to express their class, people have to make their socio-economic status visible through their lifestyle, that is, by their everyday practices and consumption patterns. Through their lifestyles, people do express both who they are (identity), who they want to be (aspirations) and who they do *not* want to be (distinction). In doing so, they relate not only to their own class but to other social classes. Namely, in their aspirations, they relate to upper classes. In distinction, they mostly relate to lower classes.

As aspirations often result in imitating upper-class lifestyles, these lifestyles are normalized over time, and the upper classes have to continuously change their lifestyles in order to be still able to distinguish themselves from the lower classes (Elias [1939] 2000). Therefore, lifestyles and associated consumption practices have been continuously changing over time and are deeply engrained into specific socio-cultural contexts – what is done and consumed by whom in which ways can only be understood when understanding social contexts' history.

In their empirical work, most inequality and lifestyle researchers have focused on Europe or North America. In this paper, we will discuss how analysis and resulting conceptualizations change, if other social contexts are integrated into the analysis. We will concentrate on *food* and *housing* in urban space, as lifestyle research has shown that in most social contexts, these are the key markers of making social class visible in everyday life. More specifically, in order to be able to better illustrate the differences between the social contexts, we will focus on urban middle-class households and illustrate how they ideal-typically relate to other social classes, how they position themselves in urban space, how they do distinction by eating and housing practices, and how this interrelates with the organization of the food market. We will do this by comparing three contrasting cities of comparable sizes – Berlin (Germany), Singapore and Nairobi (Kenya). Our analysis is based on joint empirical work in all three cities (Section 2). Note that we will argue ideal-typically in order to better shed light to the differences between the social contexts. We will start our discussion with how social class (socio-economic status) is related to other key dimensions of social inequality – namely gender, age and race and ethnicity – and how this results in family structures and housing practices in the contrasting contexts (Section 3). We will continue by arguing how social class is expressed in people's eating practices and the foods they buy and eat (Section 4), how social inequality is engrained in urban space as a whole and in specific neighborhoods (Section 5). In Section 6, we will link these two strands of argument, showing how eating and housing practices effect the organization of the food markets in urban space. We will conclude with a discussion and outlook for future research (Section 7).

2. Data and Methods

As our research aims at deconstructing existing social concepts and building new theories, cases – the cities – were selected purposefully in order to maximize contrasts between social contexts, and embeddedness in these contexts is part of our analysis. More specifically, against both the backdrop of existing sociological and geographic literature and our own continuous exploratory ethnographic field work in various project contexts, three cities of comparable size were selected: Berlin (Germany), Singapore and Nairobi (Kenya). Namely, we selected these cities because we expected them to differ concerning their urban-rural relations and their position in the world system, typical social constellations (class and gender) and their relations between places of residence in the quarters, typical food knowledge and foodways as well as the structure for the food market.

Once the cities were selected, we used literature, cartographic data, social inequality indices and historical information in order to select between 10 and 20 neighborhoods within each urban space. We tried to maximize the variation between neighborhoods concerning building structure, embeddedness in the city as well as social and ethnic diversity. Note that the distinction of neighborhoods is not as clear-cut as it seems on first sight, which is why this number needs to remain vague.

We then conducted focused ethnography in Berlin (April – July 2022), Singapore (September – October 2022) and Nairobi (November 2022) by walking the selected neighborhoods in research teams of four to almost 30 people. Larger groups were split up in smaller teams walking the neighborhood separately. Data were documented by field notes, photography and sketching and triangulated both with observations and findings of other researchers who participated in joint fieldwork and with various other data as well as with findings from our own research projects and from other researchers. In the following analysis, we will scrutinize how different social contexts effect food and housing in their relationship to lifestyles and urban space.

3. Intersection of Class and other Dimensions of Social Inequality

In all three countries, *social class* is one of the most important dimensions of social inequality. While in Kenya, social class is mostly defined by ownership – landowners versus the landless – in Germany and Singapore, class is not only defined by ownership but paid work: access to specific job categories can only be achieved by specific educational qualifications and results in distinct occupational prestige and income. Social class typically intersects with other dimensions of social inequality such as race and ethnicity, gender and age and results in specific household patterns as well as lifestyles. In addition, class demarcations and the differing lifestyles of upper, middle and lower classes form and influence urban space.

When it comes to eating and housing practices, *gender* is usually secondary to social class in all social contexts – what people eat and how they live is typically a household decision and applies to both genders equally. Within the household, couples typically practice a *gendered household division of labor*. In all three social contexts, historically, buying and preparing food (cooking) is a woman's task as the "homemaker," while being responsible for the main family income is the man's task as the "breadwinner." In all countries, in traditional households, this household division of labor is still kept up. In modern households, men also shop for groceries and cook, but they typically do the "show cooking" when guests are there, while women are responsible for everyday cooking.

Class and gender intersect via the services middle-class families make use of in order to be able to keep up their lifestyles: within the household, these services are mainly provided by women from lower classes. While in Germany, only a small share of middle-class families makes use of *domestic services*, (usually as cleaning staff, babysitters or nurses for the elderly who enter the family home just for a few hours per week), in Nairobi and Singapore, it is a common practice to employ (female) domestic servants in the household for a wider range of services, including grocery-shopping and cooking. In terms of class distinction, this means that in Germany, lower-class persons rarely have a chance of seeing a middle-class household from the inside, whereas the classes are much more mingled in the household in the other two cities.

As Graph 1a illustrates, in Germany, this results in the home being embedded into the neighborhood being embedded into the city being embedded into the nation. Each of these nested entities is a territorial space ("Territorialraum") characterized by a logic of demarcation ("Logik der Grenzziehung") (Knoblauch/ Löw 2021), as families from different social classes are segregated in residential neighborhoods, move largely within their own neighborhood for everyday activities and typically only get access to homes of their own social class (Baur et al. 2021).

In contrast, both Singapore and Nairobi middle-class households typically make use of domestic servants, so the classes are much more mingled in the household: in 2016, one in five Singapore households had a (female) domestic servant (Paul 2017: 102). These maids typically live in their employer's apartment and are labor migrants with a limited working- and residence-permit from neighboring poorer countries such as Malaysia, Indonesia or the Philippines. Nairobi middle classes, too, have a long tradition of having (female) domestic servants. Maids may either be born in Nairobi and live in an adjunct informal settlement (slum), or they may live in their employer's house and come from the employer's home village (community) (Kibel et al. 2023). As Graphs 1b and 1c illustrate, this results in the lower classes breaking up the territorial space of a middle-class home by linking it to other spaces in a space of pathways ("Bahnenraum"), following a logic of transit ("Logik der Durchquerung") by commuting these spaces, thus creating a potential spatial conflict ("Raumkonflikt") between territorial space and space of pathways (Löw/Knoblauch 2021). For class concepts this means, that middle-class lifestyles become intimately visible for the working classes (but not necessarily vice versa).

Graph 1: Family Structures and Housing

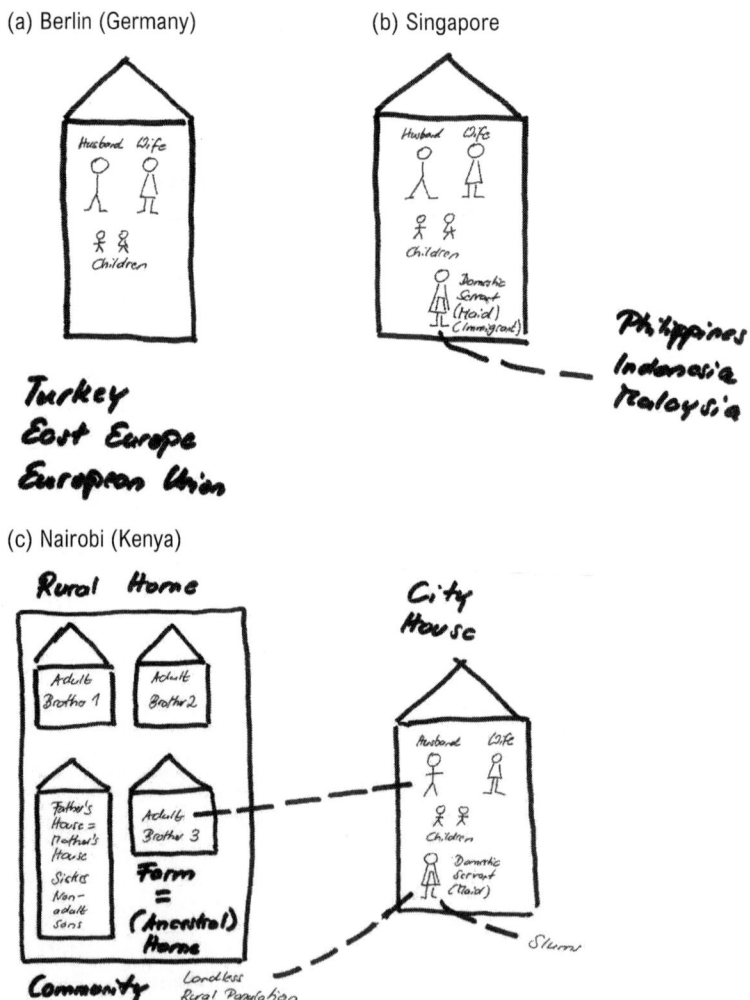

In all cities, middle-class couples' relationships to their parents and their family of origin is important to them. However, while in Germany and Singapore, becoming an adult means to move out of the parents' household (which spatially can be conceptualized as one territorial space) and founding your own household which then becomes your new home (and a new, demarcated territorial space), in Nairobi, adults remain much more entangled with their family of origin (which often still resides in a village). Accordingly, Nairobians strongly distinguish between their (city) "house" and their (ancestral) "home." In the rural areas, in landowning families, young men typically build a small house on their parents' ancestral ground as part of their passage to adulthood (Kibel et al. 2023). Thus, a rural home consists of the parents' house plus one house for each adult son. These houses are considered the actual home and kept up even if men migrate to the city. Typically, one son remains at the ancestral home, even if the other sons migrate to the city, and the migrant sons continuously return home with their families, keeping up family relationships (Kibel et al. 2023). Consequently, while in Germany and Singapore, urban and rural areas are distinct, Kenya is characterized by strong rural-urban entanglements (Graph 1c).

4. Class and Eating Practices

Lifestyle research has shown that eating practices are one of the key ways of making social class visible – not only because the lower classes have restricted budgets but also because typical tastes and ways of eating express and reinforce identity within a social group and make social class visible to outsiders (including typical body types resulting of different ways of eating, combined with work- and leisure-related activities). The upper and upper-middle classes typically follow high-brow eating practices in order to express their special tastes (Bourdieu [1979] 1984), such as eating at Michelin-rated restaurants. As the middle classes are imitating the upper classes in their aim for moving up the social ladder, the upper classes continuously have to adapt their eating habits in order to remain avantgarde (Elias [1939] 2000). As a result, upper- and middle-class eating habits can only be understood at a specific historical time and against the backdrop of the historically evolved typical ways of eating. The situation is more complicated for the working classes: Bourdieu ([1979] 1984) has shown that the working classes both have to handle restricted budgets and at the same time develop their own, typically more necessity-based tastes, so by no means can one assume that the working classes automatically imitate the middle and upper classes – on the contrary: the working classes are likely to practice their own taste cultures in order to show that they are *not* middle or upper class. So how does class unfold in the field of eating in the contrasting social contexts?

4.1 Berlin (Germany)

In many ways, how eating habits and social class are linked in Germany is very typical for Europe and resembles what Pierre Bourdieu ([1979] 1984) has shown for France in "Distinction." However, in contrast to France, due to harsh German winters, usually no locally-grown fresh produce is available between October and April, and the whole Christian calendar is organized around climate-related times of abundance and lack of food. Food and poverty have always been closely related. Only since the 1960s, there has been an abundance of food. Up to today, historical legacy lingers: traditional German cuisine is very necessity-based, aiming at filling up hard-laboring people. Meat is highly valued, as it historically was reserved for the aristocracy and the upper classes (Elias [1939] 2000). As today, meat is widely available, an average German eats 76kg meat a year, compared to 43kg in a world average (Destatis 2022). Even today, most Germans cook themselves and – if possible – eat at home. If they are outside their home – e.g. for work or education –, for lunch, the majority of Germans either bring food from home or alternatively go to a canteen or skip lunch (Mende 2022: 79–81). In Berlin (compared to the rest of Germany) prices of meals in restaurants are relatively low. Therefore, people go out to restaurants more often than in the rest of Germany. Regardless, going to a restaurant is typically reserved for special occasions in the evening or on weekends, and only a minority of Germans eat lunch in restaurants (Mende 2022: 80–88). Accordingly, when Germans think about "buying food," they think about "shopping groceries" in order to cook themselves.

Against this backdrop, the traditional working classes tend to prefer traditional German cuisine, the modern working classes fast-food and international standard cuisine such as pizza, burgers and doener kebabs. For the working classes, cooking and eating practices are mostly necessity-based: in order to save money, a typical German lower working-class woman shops for discounted food, cooks large portions at home which are portioned out, frozen and then eaten over the week. For the deprived population (and despite the cultural preference for meat), food is mostly vegetarian, with starchy staples like bread and potatoes making up for the largest share of meals (SVRV 2021: 192–194).

Historically, European aristocracy was pan-European, and in the course of the civilizing process, French cuisine became the point of orientation for European high-brow cuisine (Elias [1939] 2000). In the 1960s and 1970s, German high-brow cuisine was still dominated by French cuisine. Since then, taste-wise, the German middle and upper classes have emancipated themselves to a certain degree from French dominance of tastes: while current French high-brow eating practices are characterized by a higher tendency to gastronationalism, in Germany, the middle and upper classes tend to eat more foreign cuisines which are both considered healthier and to provide more exotic tastes than the necessity-based traditional German cuisine. As a result of typical immigrant populations' foodways, this used to be not only French but also Italian and other Southern European, Turkish and Arab cuisines. In the last decades, the palates have widened especially towards Asian cuisine. In addition, expressing an ecologically sustainable lifestyle has become fashionable in recent decades which is why vegetarian, regional and organic food has become more fashionable among the middle classes (Fülling 2022). In Berlin, increasing wealth and awareness of good food is expressed e.g. in the growing number of Michelin-rated restaurants, and the variety of available cuisines in restaurants is ever-increasing. As most Germans still cook themselves at home and being a good cook – with men doing the show-cooking (Baur/Akremi 2012) – is a way of expressing distinction, changing eating habits also have an effect on the broad variety and quality of groceries being available in middle-class neighborhoods (Fülling/Hering 2020; Baur et al. 2021).

4.2 Nairobi (Kenya)

Like Germans, Kenyans ideally eat three meals a day, the main meal being lunch. Also like in Germany, lunch typically consists of equal shares of a starch-based staples – in the case of Kenya "Ugali," a cornmeal porridge similar to polenta –, shredded cooked vegetables and proteins – which are traditionally beans. As Kenya is a much poorer country than Germany and food is still scarce, most Kenyans practice a survival economy and eating practices are much more necessity-based than in Germany – being "well-off" in Kenya means "not having to skip meals."

Like in Germany, eating meat is considered a sign of wealth. So, replacing vegetable-based proteins for meat and being able to eat a lot of meat is considered a sign of wealth which is even today reserved for the middle and upper classes. In comparison to Berlin and Singapore, in Nairobi, there is a very low number of restaurants, and *haute cuisine* has just started to develop. Restaurants aiming at higher-class audiences are mostly located at hotels, country clubs or in malls and serve international cuisine with traces of African cuisine. The point of reference for "high-brow cuisine" seems to be not French but American-British cuisine, typically serving a mixture of American-British, Italian, Indian, Chinese and African international standard dishes. However, even if they can afford it, most Kenyans do not seem to trust these restaurants, as they assume that dishes are neither freshly prepared nor stored properly and therefore seem to not only have doubts about hygiene but feel that food is less tasty. In consequence, even the middle classes consider hawker shops on the roadside as the better alternative – here women cook freshly-prepared Kenyan or sometimes Indian or Arab food, reflecting the historic immigrant populations' eating habits which have

171

been integrated into urban food. However, in most Kenyan families, a woman still cooks at home and meals are eaten at home. So, when Kenyans think about "buying food," like Germans, they think about "shopping groceries" in order to cook themselves.

4.3 Singapore

The Singaporean state has also encouraged food and eating to be a means of building cosmopolitan national identity, community and memory (Kong 2016: 220), resulting in a "foodie culture" (Kong 2016: 223–224) where hunting for the best version of a favorite dish all over town and comparing these dishes in the (social) media is a passion (Xiang Ru 2016). So being "wealthy" does not mean "having enough to eat" but "eating *good* food." In contrast to Germany and Kenya, this is not only true for the middle and upper classes but for all classes and reflects in the fact that – while it is easily possible to spend more than 1,000 dollars for dining in a restaurant – Singapore is also one of the few countries with Michelin-starred hawkers. As the upper classes can afford to downplay their social status in everyday life, it is socially acceptable for all classes to eat cheap food such as hawker food, as long as it is tasty.

While in Berlin and Nairobi, the point of reference for "good eating" is Western high-brow cuisine, Singaporean cuisine refers to Chinese cooking and eating culture. Chinese cuisine values not so much quantity but refined tastes which can be only acquired by a combination of very fresh, rare ingredients and the cook's skill in handling these ingredients. Enjoyment comes not from eating a lot but from sampling many different rare tastes. It is therefore not surprising that in the course of Singaporean history, the cuisines of the other dominant ethnic groups – namely Malay, Indian, Arab cuisine – have been integrated in everyday life and Singapore has developed its own version of fusion cuisine – Peranakan cuisine. Driven by popular culture since the 1980s as well as the increase of immigration and Singaporeans own ability to afford to travel worldwide and the resulting discovery of new foods since the 1990s, the palate of available cuisines was widened to include Japanese, South Korean, Southeast-Asian (Thai, Vietnam or Lao), "Western" and "International" (that is American-British) and European (particularly Italian, Spanish, French and Swiss) cuisine and creating new types of fusion food (Kong 2016). Like in Germany, the market for unprocessed, organic and wholesome foods has been growing in the last decades (Sinha 2016: 160). Regardless, the internal logics of Singapore's food culture can only be untangled, if Chinese food culture is understood: while in European cuisine, every person eating typically gets their own plate with their own share, in Chinese cuisine, people like eating together and sharing food from the same coming plates. In addition, many more small meals are eaten in the course of the day.

Graph 2: Hawker Centers in Singapore

Another defining feature of Singapore's foodways is that Singapore has a long tradition of street food (Graph 2) dating back to the nineteenth century (Beng Huat 2016; Kong 2016: 212). Hawker centers provide cheap food so the (female) working population does not have to cook (Sinha 2016). Going to the hawkers who cook fresh meals on demand, is a time-saving and cost-effective alternative to cooking oneself. Food courts improve hygiene and provide the hawkers with infrastructure such as electricity, fresh water and joint facilities such as toilets or dish-washing. Especially younger Singaporeans either eat all or most meals outside of the home (Sinha 2016: 171–173). So while in the rest of this paper, we will focus on grocery shopping, it is important to keep in mind that in Singapore, eating and cooking are disconnected, and therefore also grocery shopping and eating (Sinha 2016). When talking about "buying food," most Singaporeans are associating "eating out" and Singapore's food market cannot be understood without understanding hawkers. The hawkers always constitute the social and spatial heart of a neighborhood. They are typically located next to a transport hub such as a bus or subway station and are either situated in special buildings – the hawker markets – or on the lowest floor of an HDB building (basement or ground floor) (see section 5.2).

It is also important to note that while many Singaporeans of all genders, classes and age-groups like to do show-cooking, only two in five Singaporeans know how to cook (Sinha 2016: 172). Persons who actually *do* shop for groceries and cook on a daily basis are professional cooks (such as hawkers) and in home-cooking almost always women – namely, either elderly working-class women (because their family incomes are too low for going to hawkers regularly) or domestic servants (Sinha 2016).

5. The Spatial Arrangement of Social Classes in Urban Space

Not only eating practices, but also housing practices are important ways of expressing social class. When it comes to housing practices, there are three aspects of how social class is expressed in space:

1. As elaborated in section 3, social class in its intersection with other dimensions of social inequality, results in specific *housing patterns* (Graph 1). In Germany, the home resembles a territorial space. In a middle-class nuclear family, only middle-class people live in the home and no other class gets access to it. Due to the common practice of live-in domestic servants, in Singapore and Nairobi, the classes mingle in the home. In Nairobi, in addition, landowning families are bilocal and regularly commute between the ancestral home in their home village and their house in Nairobi along a space of pathways.
2. As we will show in this section, the social classes are *distributed unevenly in different neighborhoods across urban space* and, in doing so, *relate to each other* and are coupled in their everyday practices. Which social class is located where can only be understood, if one untangles a city's history starting from its formative phase. This forms an overall spatial arrangement ("Raumanordnung") of the city (Löw/Knoblauch 2021).
3. As we will show in the next session, this spatial arrangement of social classes has an effect on *how the neighborhood is structured.* We will highlight this using the example of food-retailing (Section 6)

5.1 Berlin (Germany)

While in Germany, neighborhoods are always socially mixed to a certain extent, typically, the social classes tend to segregate in different neighborhoods – this is true much more for Berlin than for other German cities, and typically, each neighborhood is dominated by a social class. Since industrialization, the home ("Wohn*ort*") and the workplace ("Arbeits*ort*") have been separated, and the working population commute between these neighborhoods, thus linking neighborhoods via their everyday practices (Kulke/Baur 2021). As a result of neighborhood segregation, middle-class Germans usually do not meet upper-class residents in everyday life at all and do not meet working-class people in their home but only outside the home – either in their neighborhood at shops, e.g. when buying food or other services which are usually provided by lower-class labor ("Einkaufs*ort*"), or at the workplace ("Arbeits*ort*"). Consequently, the neighborhood a person lives in and what the home looks like from the outside become key markers for assessing another person's social class.

Graph 3: Historical Markers of Luxury as Part of Aristocratic Lifestyle (e.g. Pfaueninsel)

Which neighborhoods are regarded as "better" or "worse" neighborhoods is strongly defined by Berlin's urban history. In its formative phase between the fifteenth and the seventeenth centuries, Berlin was founded as an aristocratic residential city (Häußermann/Kapphan 2000: 26–30). For European nobles, closeness to water – such as rivers and lakes – and access to forests and parks were signs of wealth. These were both used for fishing and hunting (providing fresh food in winter) but also for leisure and representation, as can be exemplified by the "Pfaueninsel" (an island with huge park areas within the river Havel on the Western outskirts of Berlin) (Graph 3).

Graph 4: Social Inequality and Urban Space in Berlin (Germany)

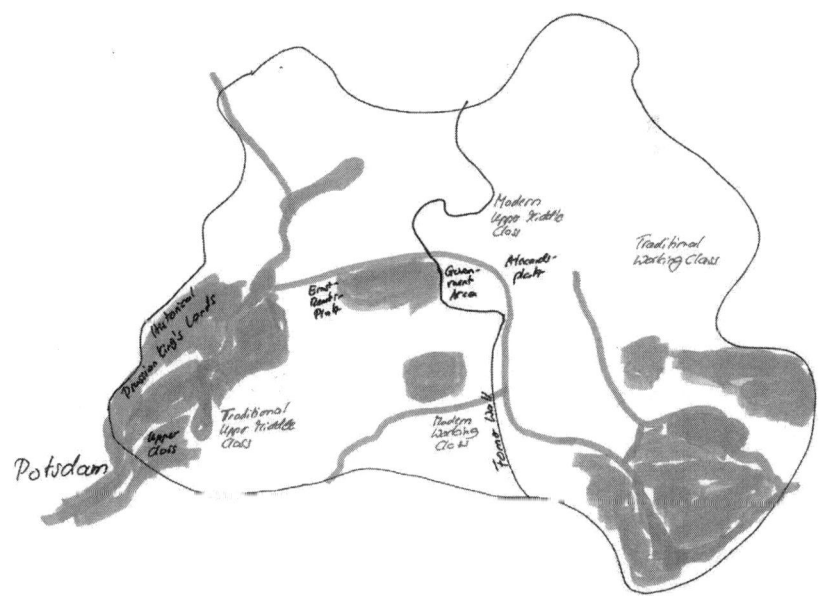

While today, most rivers, lakes, forests and parks are government-owned and open to the public, living close to them in a quiet area which is well-connected by public transport still defines a "good" neighborhood (SVRV 2021). The working classes get what the housing market leaves once the middle classes had their pick – as a rule of thumb, areas which are either located at the fringes of the city and poorly linked to public transport or are located closely to unfavorable locations such as prisons or red-light districts. Note that not only the outer limits of town belong to the fringes but also the areas close to the former wall which divides the city in two parts (Graph 4).

Graph 5: Neighborhood Types in Berlin

(a) Upper and Upper-Middle Class Housing

(b) Lower Working-Class Housing

Which social class dominantly inhabits a neighborhood can not only be inferred by its location but also by building type. As Graph 5 depicts, stand-alone housing which typically is to be found in the outskirts are a typically sign for upper- and upper-middle class housing. In contrast, high-rise buildings which were built in the course of public housing policies between the 1950s and 1990s are typically inhabited by the working classes. Between the 1840s and World War II, in

Berlin, many types of block buildings were built (Häußermann/Kapphan 2000: 31–36). Today, neighborhoods dominated by these building types can be either working- or middle-class, depending on location within town. The dominant class in a neighborhood can be inferred from its infrastructure, e.g. which restaurants and grocery shops are located there and what they offer (Section 6), so food infrastructure is not only a result of a neighborhood's class structure but also a signifier for social class.

5.2 Singapore

Singapore's urban space is of similar size as that of Berlin, but its spatial arrangement follows completely different logics (Graph 6) – again, in order to untangle them, one needs to refer to Singapore's history. During the colonial period, under British rule, a very specific urban landscape developed (Vorlaufer 2009: 110f). Both the spatial distribution and the internal structure of housing areas were not so much defined by social class but by ethnicity and the immigration regimes of these ethnic groups. Network-migration to people of the same family or people coming from the same region or business was the most important factor for choice of neighborhood: the British typically were either working in administration, as officers in the military or in the management of trading/transportation companies. Their housing areas, mostly with single houses in European style, were connected to these activities and up to today form Singapore upper-class areas, defined by typical colonial residencies (Graph 7a). Chinese immigrants worked mainly in trade, retailing and handicraft. Most historic Chinese houses were constructed next to the Singapore River in "Chinatown," which was the most important port area at this time and opened the possibility to practice their economic activities. Typically, Chinese settlers constructed "shop houses" (Graph 7b) which had two floors, the ground floor being used for the commercial activities and the upper floor for living. All houses were attached to each other having only a small front to the street and very long extension to the back mostly including one or two backyards. The building characteristics with shop houses were more or less the same in the areas of the Hindu Indian and Muslim Malay and Arab groups. Indians typically immigrated to Singapore in connection with British colonial activities – such as working in administration and military – or in order to run their own business in retail and services. Their housing area was North-East of the colonial center in "Little India." The Malay settled further to the East in Kampong Glem. As a result of these historic settlement patterns, at the time of the independence of Singapore from British Empire in 1963, the urban hosing landscape was characterized by a distinctive separation of the ethnic-religious groups.

Graph 6: Social Inequality and Urban Space in Singapore (Modern Structure)

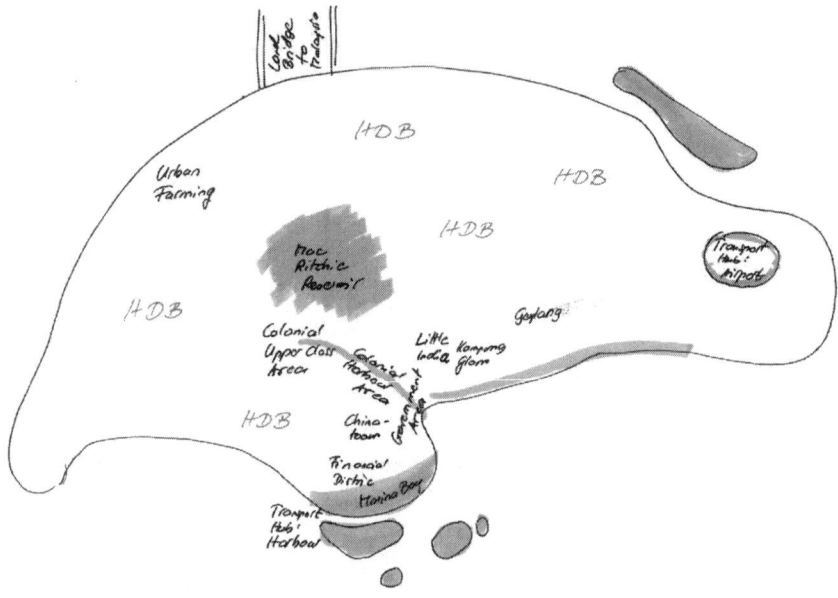

Graph 7: Colonial Singapore Housing

(a) Upper-Class Colonial Housing

(b) Shophouses

Since 1957, Singapore's population grew rapidly from 1.5 Mio. to 5.7 Mio. in 2020 (SHDB 2022), resulting in a dire need for new housing. After independence in 1965, both the Singapore government's urban development policy and fast economic growth induced a total change of the urban housing landscape and its relation to class (Kiese 2017). In the shop house areas, hygienic conditions were horrid: these areas were not only heavily overpopulated but there was also no piped water, no waste water canalization and hardly any electricity available. Therefore, there was an urgent need to build additional housing areas to accommodate the growing population and to resettle people from the shop house areas. Already in 1960, the Singapore Housing Development Board (SHDB) was established to construct new neighborhoods and new towns for the people. Since then, newly constructed neighborhoods have been highly uniform, resembling Socialist social housing but aiming at all classes.

Graph 8: Singapore Housing Development Board Flats (HDB Flats)

Graph 9: Social Structure Represented in HDB Housing in Singapore

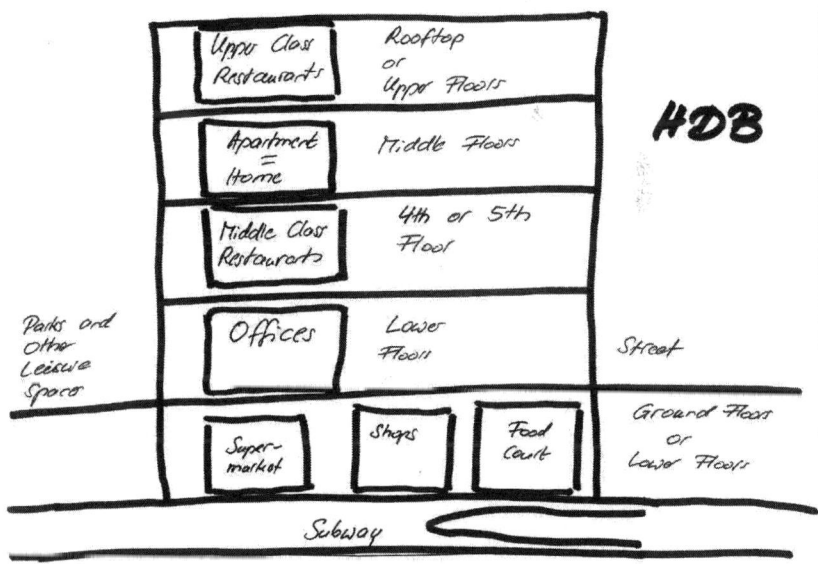

179

In contrast to Berlin, in Singapore, there are no urban fringes – HDB neighborhoods (Graph 8) are always linked to town by a central transport hub (bus or metro station), resulting in travel distance to the center being more or less the same, regardless where you live. Early HDB neighborhoods had several multi-storey buildings which surrounded a central service area with retail stores, hawker centers, medical services, school and library. Several of the neighborhoods then formed a new town with a central area with more advanced services in retailing, gastronomy, medical service or education. In current HDB neighborhoods (Graph 9), all these functions are integrated into the one high-rise building, which means that urban space cannot be understood without thinking three-dimensional: residents arrive at ground level or underground via bus or subway. When entering, they can shop for groceries and other goods and grab some food at a hawker center or food court which are also located underground or on the ground floor. Also on the ground levels, there are parks and community centers for leisure activities. Offices are typically located at the lower levels, followed by mid-range restaurants (typically on the 4th or 5th floors), the apartments on the upper floors and then up-range restaurants and bars on the rooftop and upper floors, typically nested in a lush urban garden with grand harbor or city views. Everything is accessible to everyone, e.g. by many walkways which link diverse places.

However, the most important element of HDB housing policy is to have within the housing blocks a multi-ethnic, multi-religious and multi-class mixture representing the whole range of Singaporean society ("Building helps to promote social cohesion and racial harmony, strengthen family ties, care for the needs of the elderly, singles and low-income families"). In the same house, and even on the same floor, small apartments for low-income households and large apartments for higher-income households have been constructed and then sold. This housing policy has been highly successful: while in 1957, 8% of Singaporeans lived in HDB housing, in 2020, about 80 % of the population of Singapore is living in these public housing estates (SHDB 2022). With this share, Singapore is probably the city with the strongest socio-economic and religious-ethnic mixture in housing areas in the world. Ironically, the politics of social mixing do not diminish social class differences but instead make them more visible. As a result, fewer markers are needed to demarcate them from each other, namely the size of the apartment, owning a car and being to afford fine dining once in a while.

Furthermore, this mixing process is not linear: as a result of globalization, the rise of a very-high-income class of Singaporeans and the immigration of highly-qualified high-income expatriates with a permanent residence permit, first trends of re-separation and re-segregation can also be observed since the 1990s when public housing peaked at 86% of Singaporeans living in public housing. Since then, this share has slightly reduced. Today, residents with very high incomes prefer to live in private condominiums. These buildings are typically located either in the colonial upper-class areas or renovated former shop house areas (Graph 7) – making the historical town center a highly gentrified area – or they are located at highly esteemed locations close to the sea or parks, designed by well-known architects and offering private facilities like tropical gardens, swimming pools, tennis-courts or sports-centers to a very homogenous social group. They are also typically gated in very subtle ways and guarded by security personnel, excluding people which do not belong to this social-class (Graph 10). As they are often located next to HDB housing and offer open views inside, this makes upper-class lifestyle extremely visible. Therefore, they do not break with the Singapore tradition of inhabitants of all classes and ethnicities intermingling in neighborhoods. This makes Singapore's spatial arrangement of urban space very distinct from Berlin.

Graph 10: Modern High-Brow Housing in Singapore

5.3 Nairobi (Kenya)

Nairobi's arrangement of housing and social class in urban space is totally different from that of Berlin and Singapore and very much represents typical characteristics of extremely fast-growing cities in the Global South. Founded in

1899 as the headquarter location of the Uganda Railway, in 1906, Nairobi became the administrative center of the British colonial protectorate. At that time, only 11,500 people were living there. Until 1960, the population had grown to 251,000 people. Since then – with demographic change, economic development and rural-urban migration – the number of inhabitants almost exploded. Within sixty years, Nairobi's population increased to 4.4 Mio. inhabitants in 2019 (KNBS 2022). Housing development in this period was characterized by a combination of the urban government's attempt to plan and to steer Nairobi's spatial development and of the rapid unplanned spreading of informal settlements. As a result, urban housing areas extremely diverge in the structure of both their buildings and their inhabitants' socio-economic status. In terms of spatial distribution, there is no larger part of the town hosting only one social class – except perhaps Parklands in the North-West, which is dominated by the upper-middle and upper class. Instead, urban space is characterized by a kind of symbiosis between neighborhoods of the urban poor and neighborhoods of the middle classes.

In terms of construction and accessibility, the urban landscape of the upper-middle and the upper class is very different from other neighborhoods: landowning families living in Nairobi do not only typically have ties to their ancestral homelands, but also, the way these homelands are built reflects in the way urban neighborhoods are constructed (Graph 11a): single houses with large gardens or modern apartment complexes with joint facilities (like swimming pools, tennis courts) are characteristic for these areas. There is a lot of green space. All these areas are gated communities – there are fences, guards and video control, and access is only permitted to the inhabitants and their guests. In contrast to Singapore, it is usually not possible to see inside from the outside.

Most middle-class housing areas (Graph 11b) have a systematic street network and technical infrastructure (piped water, waste-water disposal, electricity). These areas vary widely concerning the inhabitant's job and incomes, resulting in a kind of socio-economic mixture of lower-middle and middle class.

In a way, both middle- and upper-class housing is untypical for Nairobi, as today, more than half of Nairobi's population is living in informal settlements. Kibera for example – with more than 200,000 inhabitants – is often addressed as the largest squatter area in Sub-Sahara Africa. Squatter settlements have been informally built by the inhabitants themselves using simple materials like wood, tin or bricks, and they are mostly located on land which was not planned for housing (Graph 11c). These areas are highly vulnerable to landslides or flooding during heavy rainfall; and they don't have water supply, waste-water disposal or garbage collection. In combination with the use of wood or charcoal for cooking, they are usually heavily polluted. Most of the inhabitants form the socio-economic class of the urban poor and are trapped in the so-called "survival economy" which means that each day, they only earn enough money to cover their basic food-needs.

Graph 11: Neighborhood Types in Nairobi

(a) Upper-Class Housing (Kasarani)

(b) Middle-Class Housing (Kasarani)

(c) Squatter Settlement (Gituamba)

Concerning spatial distribution, there is no concentration of squatter settlements in any part of Nairobi – instead, informal settlements can be found in almost all areas of the town. One factor for the actual site of constructing a slum is the availability of land not designated for housing, but in addition, there is a specific economic symbiosis for being located near the middle-class or upper-middle-class areas: typically, to each upper- and middle-class neighborhood, a slum is attached where the people serving the better-off neighborhood live in. Via the upper- and middle-class neighborhoods, this spatial arrangement of linked neighborhoods is also linked to the rural community of origin (Graph 12). The better-off neighborhoods offer job opportunities for the urban poor like working as maid, gardener or guard in higher-income households or running informal economic activities in the street like hawker, shoe-shine, motor-cycle-taxi or garbage collector.

Graph 12: Social Inequality and Urban Space in Nairobi (Kenya)

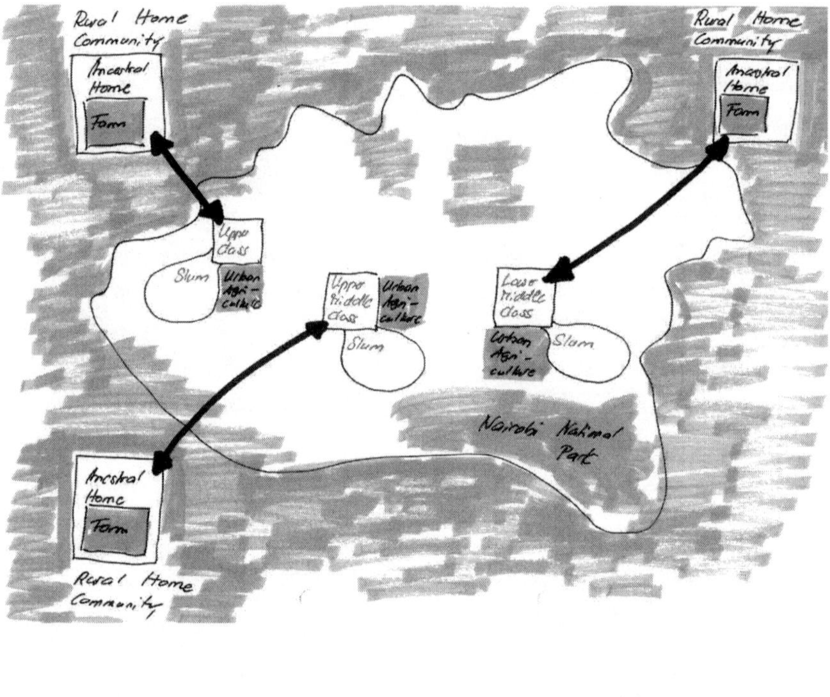

6. The Effect of Neighborhood Structure on the Food Market

Social class not only influences in which neighborhood people end up, but also how a neighborhood is internally structured. We will illustrate this by showing how the organization of the food market and food retailing is defined by the spatial arrangement of class in a neighborhood. Our elaboration will highlight, that even in a globalizing world with tendencies of standardizing cities, the local context conditions are forming unique and very specific local neighborhood and food market systems – and at the same time, the food market makes social class visible not only to inhabitants but also to passersby, thus visualizing lifestyles. Note that in the following discussion, we will only focus on consumers shopping for groceries which they plan to cook at home.

6.1 Berlin (Germany)

As discussed, in Berlin, each neighborhood is a territorial space of its own dominated by a specific social class. The territorial logic is enforced as most people will buy their groceries at the nearest center, usually within a distance of 2 km of their home. "Shopping for food" typically means "grocery shopping" either in a discounter, a supermarket, an organic/delicatessen/migrant grocery shop or at a weekly fresh market (Baur et al. 2021). Therefore, from the point of view of retailers, competition is limited to the other retailers in the neighborhood, but also, all market segments can be found in all neighborhoods (Graph 13).

Graph 13: The Food Market in Urban Space in Berlin (Germany)

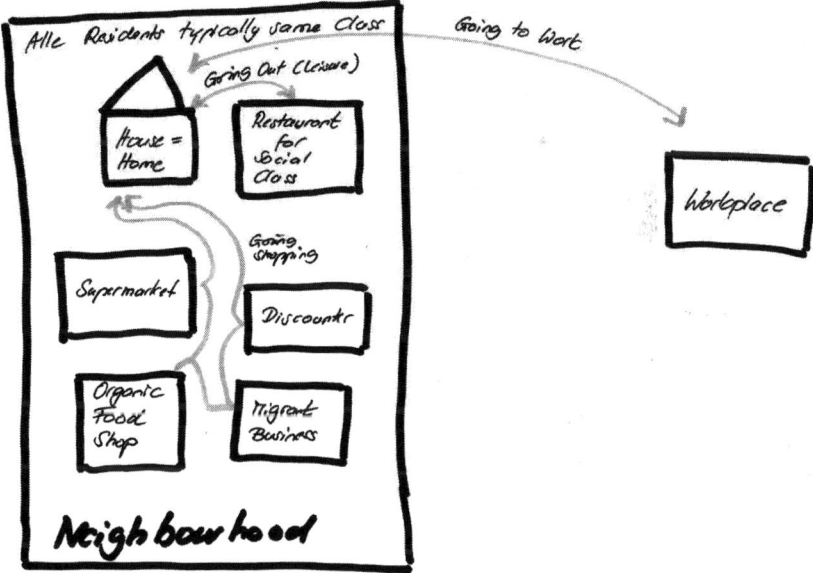

the same retailer offers vary widely between neighborhoods, and the supply structures for fresh produce and the demand patterns differ between these territorial units (Baur et al. 2021). In working-class neighborhoods, there are more conventionally-produced, low-priced and possibly migrant-oriented offers; consumers have a strong orientation towards low-priced products and their knowledge about different ways/production methods or the spatial origin of the products is limited. Looking at the retail formats, lower-income areas have a higher share of discounter offering cheaper products, and even in the same chain-store supermarket (e.g. EDEKA, REWE), the supply of articles is less diversified than in middle-class areas. In upper-middle-class neighborhoods, the variety of offers in the same stores is much greater (parallel offer of conventional, organic, fair trade, regional fresh food and vegetables) and there are even some retail formats – like supermarkets only offering organic products – which are not located in low income areas (Fülling/Hering 2020; Baur et al. 2021).

6.2 Singapore

If Singaporeans shop for groceries to cook at home, they can choose from a variety of retailing formats ranging from wet and fresh markets, state-run supermarkets such as FairPrice, privately-owned mid-range supermarkets and luxury organic grocery shops such as Little Farms. Corresponding with the SDHB politics of social mixing, all retail formats can be found in almost all housing estates, the only exception being the luxury shops which can only be found in gentrified neighborhoods. Most important for the supply systems for the daily shopping of consumers are market stalls in indoor markets, supermarkets of different sizes and hawkers.

Both wet markets and supermarkets offer a broad variety of products and are typically located directly next to hawker markets. Vendors at fresh markets play an important role in organizing the food market. They seem to represent an interface of knowledge transfer between consumption and supply (and possibly production) and they fulfill a social function in daily communication. There are typically long-term stable interrelations between consumers and market vendors. In supermarkets, the information transfer between buyer and seller is limited and only little information can be found at the products (e.g. country of origin).

6.3 Nairobi (Kenya)

Nairobi's urban food market is quite distinct. While in Germany and Singapore, all social classes will buy in *supermarkets* – only different things, depending on what they can afford –, in Nairobi, going to the supermarket in itself is a sign of a high-brow lifestyle, reserved to the upper-middle and upper classes. Therefore, while in Germany and Singapore, all types of supermarkets can be found in all neighborhoods, in Nairobi, supermarkets (Graph 14a) are typically located in upper-middle class and upper-class areas either within the neighborhood or in a mall closely located to the neighborhood (Sonntag/Kulke 2021). As both upper-middle-class and upper-class neighborhoods and malls are gated, the lower classes are effectively kept out. Supermarkets also have a key function in distinguishing social status within the upper-middle and upper classes: depending on how "fancy" or "down-to-earth" the supermarket is, one can tell, how well-off the specific neighborhood is. Supermarkets are characterized by a distinct variety of fresh products, and sometimes even specialized green-grocery shops can be found in malls. Both formats serving the broad variety in consumption preferences of high-income households.

Graph 14: Product Ranges in Different Retailing Formats

(a) Fresh Food Section in a Supermarket

(b) Wet Market

(c) Market Stall in Middle-Class Neighborhood

(d) Market Stall in Squatter Settlement

In contrast, in lower-middle-class neighborhoods, there is a parallelism of more diverse market stalls with fresh produce (Graphs 14c), wet markets (Graphs 14b), and individual small self-service shops with mostly packaged durable goods.

Graph 15: The Food Market in Urban Space in Nairobi (Kenya)

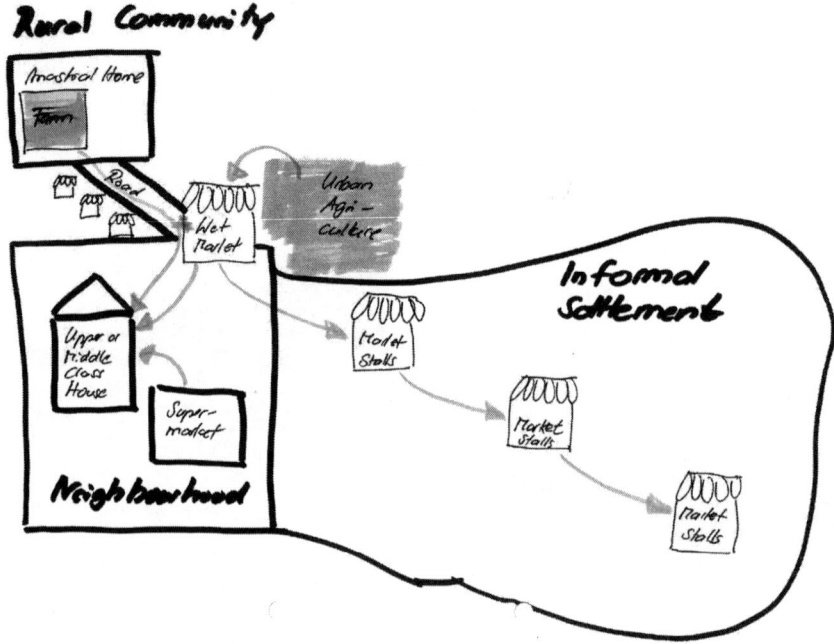

Another important feature of how the food market is entangled with Nairobi's urban structure (Graph 15) is that *rural-urban connections* between the ancestral homes and the urban neighborhoods define the food markets in several ways:

1. Commodity chains are typically organized along this space of pathways (Sonntag 2021).
2. Anyone coming back from the rural areas – including the working classes – tries to stock up in fresh produce on the way. The result is that the way back to town resembles a pearl chain of small stalls farmers have set up to sell their produce to Nairobians returning home (Graph 16a).
3. Many landowning families – when visiting their home on weekends and holidays – also stock up on fresh produce from their farms. This results in the higher classes not only having access to much fresher food than the lower classes but also a chance of additional revenue because they can sell

off excess food on the wet markets. The lower classes at the end of the chain generally have to pay more for lower-quality food, therefore reproducing social class and reinforcing the specific intersection of the neighborhoods that is so specific for Nairobi's spatial arrangement. As a result, in squatter settlements, food supply is extremely limited and food is sold in very small quantities (Graph 14d). Small food stalls at the roadside are offering only one or a few types of vegetables and they are selling them in small bundles. This is also because urban-poor shoppers only can buy what they can afford at that day.

Graph 16: Rural-Urban Entanglements in Nairobi

(a) Shop on the Road-Side from Rural Home to Nairobi

(b) Urban Agriculture

(c) Urban Gardening in Squatter Settlement

(d) Hydroponic Farm in Squatter Settlement

The rural-urban divide is further broken up by many Nairobians setting up *urban agriculture* within the city boundaries (Kulke et al. 2022), ranging from larger fields on unused lands (Graph 16b) and smaller gardens in backyards (Graph 16c) to hydroponic plants (Graph 16d). Although food produced in

Nairobi's urban gardens is typically grown on contaminated soils with poor water in contrast to supermarkets which have a strictly-controlled supply chain and therefore offer much healthier food, customers (amazingly) believe that locally-produced fresh produce is fresher and healthier than in supermarkets. In consequence, while supermarkets (Graph 14a) have the function of demarcating different higher-class neighborhoods from each other and from the informal settlements, even many middle-class Nairobians do not buy in the supermarkets but prefer to acquire their fresh produce on wet markets (Graph 14b) or food stalls (Graph 14c) – either by buying it themselves or by sending their maids to the market.

7. Conclusion and Outlook

Referring to Jörg Blasius' work, we have shown that even in contrasting cities such as Berlin, Nairobi and Singapore, eating and housing are important ways of expressing social inequality in everyday life. However, our analysis has also revealed, that the ways social class intersects with other dimensions of social inequality, and the ways the social classes are entangled with each other and related to each other in urban space vary very much depending on social contexts. These entanglements can only be understood, if a social context's history is taken into account. This in turn results in very different structures of the urban food markets. For future research, this has at least five consequences:

1. Concerning sociology of knowledge, while actors know their own local social context, this knowledge is often implicit in everyday life but needs to be disclosed in order to understand the semantics of lifestyles – most of this implicit knowledge further needs to be untangled cross-culturally.
2. Concerning lifestyle research, scrutinizing this knowledge even superficially reveals that many of the concepts and indicators used in European lifestyle research have to be reexamined when comparing Europe to other world regions.
3. Regarding social science methodology, due to different spatial logics, there is a need for scrutinizing what this means especially for quantitative analysis.
4. Concerning sociology of space, our analysis reveals that European sociology tends to think about space in a very territorial way – for example, homes are nested in neighborhoods which are nested in cities which are nested in nations and can be clearly demarcated from other territories of the same scale. In contrast, in Nairobi and Singapore, different spatial forms seem to be entwined, resulting in conflicts between spatial logics. This in turn results in the question about how these spatial logics are resolved in everyday interaction.

5. Our analysis points to an issue that has been known for a long time but is still mostly ignored in research practice of economic sociology, economic geography and sociology of consumption: that in commodity chains, everyday life(styles), including consumption, are linked to production and relate to and influence each other. The meeting point is retailing which is always embedded into the urban space which is why it seems worthwhile to analyze this entanglement in more detail.

All in all, while Jörg Blasius can look back on a long and productive career, analyses building on and relating to his work reveal that there is still a lot to do for the research community as a whole.

Acknowledgements

This publication was jointly funded by the German Research Foundation (DFG) – as part of the projects "Apples and Flowers. Effects of Pandemics on the (Re-)Organization of Commodity Chains for Fresh Agricultural Products" and "Knowledge and Goods II: Communicative Action of Consumer and Intermediaries" (A03) in the Collaborative Research Centre "Re-Figuration of Spaces" (CRC 1265) –, the Berlin University Alliance (BUA) – as part of the project "Urban Agriculture in Nairobi" – and by the German Federal Ministry for Economic Cooperation (BMZ) via the German Academic Exchange Service (DAAD) – as part of the "Global Center of Spatial Methods for Urban Sustainability" (GCSMUS).

References

Baur, Nina/Akremi, Leila (2012): Lebensstile und Geschlecht. In: Rössel, Jörg/Otte, Gunnar (eds.): Lebensstilforschung. Wiesbaden: VS-Verlag, pp. 269–294.
Baur, Nina/Kulke, Elmar/Hering, Linda/ Fülling, Julia (2021); Dynamics of Polycontexturalization in Commodity Chains. In sozialraum.de 13, 1. https://www.sozialraum.de/dynamics-of-polycontexturalization-in-commodity-chains.php [Access: 07.02.2021].
Beng Huat, Chua (2016): Taking the Street Out of Street Food. In: Kong, Lily/Sinha, Vineeta (eds): Food, Foodways and Foodscapes. Singapore: World Scientific Publishing, pp. 23–40.
Bourdieu, Pierre ([1979] 1984). Distinction. New York: Routledge.
Destatis (Statistisches Bundesamt) (2022): Globale Tierhaltung, Fleischproduktion und Fleischkonsum. https://www.destatis.de/DE/Themen/Laender-Regionen/Internationales/Thema/landwirtschaft-fischerei/tierhaltung-fleischkonsum/_inhalt.html [Access: 31.01.2023].
Elias, Norbert ([1939] 2000): The Civilizing Process. Oxford: Blackwell.
Fülling, Julia (2022): Wonach schmeckt Herkunft? Würzburg: Würzburg University Press.

Fülling, Julia/Hering, Linda (2020): Markt – Quartier – Milieu. https://www.geographie.hu-berlin.de/de/institut/publikationsreihen/arbeitsberichte/download/Arbeitsbericht_197 [Access: 30.12.2022].

Groh-Samberg, Olaf/Büchler, Theresa/Gerlitz, Jean-Yves (2020): Soziale Lagen in multidimensionaler Längsschnittbetrachtung. Berlin: BMAS.

Häußermann, Hartmut/Kapphan, Andreas (2000): Berlin: von der geteilten zu der gespaltenen Stadt? Sozialräumlicher Wandel seit 1990. Opladen: Leske + Budrich

Kibel, Jochen/Kitata, Makau/Baur, Nina (2023): Refigured Homes. In: Ernst, Stefanie/Dahl, Valerie (eds.): Dynamics of Gender Relations. Process-Sociological Perspectives. In Review.

Kiese, Matthias (2017): Singapur. In: Geographische Rundschau, 69, 1, pp. 47–53.

KNBS (Kenya National Bureau of Statistics) (2022): Population and Housing Census Kenya. https://www.knbs.or.ke/2019-kenya-population-and-housing-census-results/ [Access: 30.12.2022].

Kong, Lily (2016): From Sushi in Singapore to Laksa in London. In: Kong, Lily/Sinha, Vineeta (eds): Food, Foodways and Foodscapes. Singapore: World Scientific Publishing, pp. 207–242.

Kulke, Elmar/Baur, Nina (2021): Spatial Transformations and Spatio-Temporal Coupling. In: Million, Angela/Haid, Christian/Castillo Ulloa, Ignacio/Baur, Nina (Hg.): Spatial Transformations. London/New York: Routledge, pp. 151–166.

Kulke, Elmar/Sonntag, Christian/Suwala, Lech/Baur, Nina (2022): Urbane Landwirtschaft in Nairobi. https://EconPapers.repec.org/RePEc:zbw:esrepo:266457 [Access: 05.02.2023]

Löw, Martina/Knoblauch, Hubert (2021): Raumfiguren, Raumkulturen und die Refiguration von Räumen. In: Löw, Martina/Sayman, Volkan/Schwerer, Jona/Wolf, Hannah (ed.): Am Ende der Globalisierung. Bielefeld: transcript, pp. 25–28.

Mende, Julia von (2022): Zwischen Küche und Stadt. Bielefeld: transcript.

Müller-Schneider, Thomas (1996) Wandel der Milieulandschaft in Deutschland. In: ZfS, 25, 3, pp. 190–206.

Paul, Anju Mary (2017): Multinational Maids. Cambridge: Cambridge University Press.

Schulze, Gerhard (1992): Die Erlebnisgesellschaft. Frankfurt am Main: Campus.

SHDB (Singapore Housing Development Board) (2022): Infoweb. https://www.hdb.gov.sg/cs/infoweb/homepage [Access: 30.12.2022].

Sinha, Vineeta (2016): Mapping Singapore's Culinary Landscape. In: Kong, Lily/Sinha, Vineeta (eds): Food, Foodways and Foodscapes. Singapore: World Scientific Publishing, pp. 159–184.

Sonntag, Christian (2021): Wie kommen Obst und Gemüse in Supermärkte im Globalen Süden? Würzburg: Würzburg University Press.

Sonntag, Christian/Kulke, Elmar (2021): The Expansion of Supermarkets and the Establishment of Delivery Systems and Intermediaries for Fresh Food and Vegetables. In: Die Erde, 152, 3, pp. 166–183.

SVRV (2021). Gutachten zur Lage der Verbraucherinnen und Verbraucher 2021. Berlin: SVRV. https://www.svr-verbraucherfragen.de/wp-content/uploads/SVRV_Gutachten_2020.pdf [Zugriff 31.01.2023].

Veblen, Thorstein (1899): The Theory of the Leisure Class. New York: The Macmillan Company.
Vorlaufer, Karl (2009): Südostasien. Darmstadt: Wissenschaftliche Buchgesellschaft.
Xiang Ru, Amy Tan (2016): Bloggers, Critics and Photographers in the Mediation of Food Consumption. In: Kong, Lily/Sinha, Vineeta (eds): Food, Foodways and Foodscapes. Singapore: World Scientific Publishing, pp. 185–296.

Part IV: Data quality and statistical education

Sociological perspectives on teaching statistics in sociology – disciplinary complexities, interdisciplinary challenges and data worlds

Rainer Diaz-Bone

> "As sociologists, we problematize every subject we teach – from race to religion and from the family to world-economy – every subject except for statistical methods." (Moran 2005: 264)

1 Introduction

Statistics is an important tool for sociological research. The institutionalization of sociological study programs since the second part of the twentieth century and sociology as a profession would not have been established without statistical training and quantitative analyses of social data. Monitoring and representing society, studying its structures and dynamics can only be done by applying computational skills, modern data analysis software and statistical approaches. The establishment of survey programs and durable data infrastructures, the rise of datafication and the Internet have facilitated sociological research and support conditions for teaching in statistics (especially because of open access to data, advanced statistical open access software like R and cheaper as well as more powerful computer technology). But there are other changes in sociology, in statistics, in the field of sciences as well as in society itself, which question this advantageous picture for the relation of sociology and statistics as well as the conditions and perspectives for statistics training in sociology. It is not a new experience for sociology to be challenged or to be questioned, but as a discipline, sociology has always worked out its specific contribution to developing and evaluating social policies, to analyze society in all its dimensions and to exercise social critique based on scientific analysis, too. Since its professionalization, i.e. since Paul Lazarsfeld's seminal works, no other discipline in the field of social sciences has elaborated such a broad set of methods and methodological strategies as sociology. Still, sociologists are the ones who are most important in developing and proliferating empirical social research methods (as well as theoretical approaches) suitable for other social sciences as well (see Baur and Blasius 2022; Diaz-Bone and Weischer 2015). But in the case of statistics it is different, because it is provided mainly from other disciplines to sociology and has to be integrated in sociological frameworks, paradigms, research and ways of teaching.

In this contribution, some contemporary developments and problems for teaching statistics in sociology will be sketched. It will be argued that sociology's situation is specific because of the increasing complexity of its internal organization and upcoming challenges from its contexts – scientific ones as well as societal ones. All of these have an impact on teaching statistics, to which sociology has to find an answer. It will be argued that empirical sociology and sociological research cannot be integrated on the basis of theory or methods or even statistics alone but has to work on its plurality of "theory/methods packages" (Clarke et al. 2018), which are coherent ways to align theoretical concepts, statistical tools and methodological principles and epistemological values to employ both. Teaching statistics in sociology, therefore, has to introduce and to focus on this co-existing plurality of theory/methods packages. Also, the focus has to broaden, which is not to be restricted to class room situations or pedagogical issues only, but to bring in an institutionalist perspective and to regard the implementation of statistical teaching from the standpoint of a sociology of social research.[1] The institutional approach to develop a sociology of social research is "convention theory", which regards conventions as "institutional logics" in data worlds (Diaz-Bone 2018; Diaz-Bone and Larquier 2022).[2]

2 Facing increasing and multiple complexities

There are many reasons, why the statistical teaching has become more and more complex.

First, the evolution of statistical procedures has been ongoing for a long time and resulted in a huge landscape of statistical tools and logics. Many of these innovations in the last decades are of relevance for sociology, because they approach the specific situation of social data; such as procedures for categorical

---

1 See for research on teaching statistics the journals "American Statistician", "Teaching Statistics", "Teaching Sociology" or "Journal of Statistics and Data Science Education". For educational analyses and reflections on teaching statistics in schools and universities see the contributions in Ben-Zvi and Makar (2016) and Ben-Zvi et al. (2018).
2 This approach was developed in France and is named "économie des conventions", "economics of convention" or "sociology of conventions", too.

data analysis, network analysis, multi-level modeling, event history analysis and many others. Statistics can be employed to explore data structures, to test hypotheses or to visualize complex data by more and more data analysis software, which require comprehensive data wrangling skills and coding capabilities. Data resources, data types, data volumes, data archives and data generating devices have multiplied, too. What has not enlarged and evolved is students' knowledge in basic mathematics or computational skills, which are both needed to study and apply statistics. A substantial part of beginners is barely able neither motivated to approach statistics lessons, because the decision to study social sciences was taken without a substantial knowledge about the importance of social research skills and their central role for the professionalization of these scientific disciplines. Most social sciences bachelor programs have included obligatory courses in statistics in the "methods" section. But it is an open question whether and how BA programs manage to let students experience the use and necessity of methods and statistics, in order to apply them skillfully in empirical analysis and in answering research questions. To succeed, students and researchers have to adopt knowledge about research designs, explanatory logics and conceptual networks, i.e. theories on forming and answering research questions.

Here, a second complexity comes to the fore. Sociology is characterized by a plurality of explanatory logics: methodological individualism, methodological holism, methodological situationalism (and one could add methodological relationalism), which employ different kinds of entities as explanatory principles. In addition, some of them distinguish different kinds of levels (micro, meso and macro level) and intermediating mechanisms. These explanatory logics are linked to sociological theories, because these logics need to indicate on one side entities and principles, which cause, impact or interlock and interact, i.e. have agency, and on the other side these logics have to indicate entities and outcomes, which do not have agency on their own (i.e which are emergent realities or effects). Due to the differing logics, the theories to which they are linked differ as well.

Now, a third complexity appears. Sociology is an open, multi-paradigmatic discipline. Thomas Kuhn (1962) coined the notion of "paradigm" and portrayed scientific progress not as the accumulation of knowledge, but as a sequence of dominating paradigms. However, empirically, almost all disciplines have different paradigms that coexist at the same time. In most sciences, one paradigm is dominating and the others are marginalized. But in sociology, there are plenty of influential paradigms, many paradigms are marginal, no single one dominates the discipline. While some paradigms are linked to interpretive (in contrast to standardized) research methods and designs, there are several paradigms which offer different explanatory principles and are linked to different explanatory strategies and all of them employ statistical tools – but different ones and in differing ways. In sociology, there is a coexisting plurality of statistical cultures, not only because of the multitude of statistical procedures and approaches, but also because of the many coexisting explanatory logics and paradigms, which exist in different combinations in sociology as ways of statistical analysis.

To demonstrate this specialty, one can compare sociology with other disciplines, e.g. economics. In economics, neoclassical economic theory (rational choice theory), methodological individualism and econometrics (regression-based statistics, i.e. generalized linear modeling) form the dominant combination. In this case, there is an inner affinity and correspondence between the ontological assumptions of neoclassical economic theory, its propositions about agency, the structure and logics of explanations (causal analysis of actors' decisions) and statistical approach (linear and unidirectional modeling of causes and effects on the level of individuals). This is what Adele Clarke et al. (2018) have called a "theory/method package". On the basis of this package, economics can achieve a high degree of paradigmatic unity by shared concepts, methods and compatible statistical tools. Because all other economic approaches and paradigms have been marginalized in the last decades, there is only one theory/method package in economics, which makes it a poor project in terms of philosophical discourse and competition between scientific paradigms. But it simplifies the situation for students' training enormously: economics is not too much occupied with issues such as data production; rational choice theory is based on only a few psychological statements; generalized linear models are the "same of a feather", i.e. they are all regression techniques and derivations (to model the relation between one dependent and a set of independent variables), and many aspects under study can be expressed in metric scales and units (e.g. economic values in currency units). Mathematical skills are a requirement in economics, but except for this, there is not too much cognitive complexity and students can be trained in straight and lean ways as well as in time available to reach the issue at stake (learning the required combination of skills and knowledges) and to start economic research.

3 External challenges and data worlds

The complexities mentioned in sociology are additionally complicated by interdisciplinary developments and changes outside academia.

3.1 Economics as example?

Economics, which is itself strongly influenced by experimental and quantitative psychology, has been increasing its influence on other social sciences. One reason is the strong internal coherence of economics and its image as a "science", because of its statistical "rigor" and the level of mathematical sophistication. Especially political sciences have come under the influence of economics, applying more and more economists' theories, models and ways of argumentation. In difference to political science, sociology is forearmed against the threat of a "takeover by economists", which is one of the advantages of the multi-paradigmatic structure of sociology as a discipline. But the appeal of disciplines, which aim to impress by their status as a "real science" (like economics) is exerting influence on some fractions in sociology, resulting in inner-disciplinary tensions and disputes (see for the German case Schmitz et al. 2019). Another result is that some sociology departments offer their students a program of methods and statistics that corresponds to the theory/method package of economics described above. This training normally ends with multiple regression analysis and does not include the teaching of statistical tools which approach the different situation of sociologists with regard to the most important measurement levels, i.e. multivariate statistics for data with mainly nominal and ordinal measurement level. Also, sociologists are in need of statistical education that also includes exploratory approaches, which is of little interest for economists.

3.2 Data science and computational social science

Statistics as a scientific field is facing new trends such as "big data analytics", "data mining", "data science" or – more related to sociology – "computational social science". These trends also have their origin in the rise of the datafication and computerization, the equipment of everyday technologies with digital sensors (mobile phones and apps, wearables, smart speaker assistants, modern highly computerized cars, smart homes and smart offices and so on) and the technical facilities of the Internet to link and store all the data. When inspecting books devoted to these trends and domains, it becomes apparent that in social sciences, the main element are already established statistics, together with topics formerly known as data analysis, software programming etc. (Alvarez 2016; Cioffi-Revilla 2017; Edelmann et al. 2020). In fact, some new contributions are included (like decision trees, machine learning) because data science has its origins also in other fields, such as engineering and computer science, too (Kelleher and Tierney 2018).[3] But in general, data science and computational social science are about statistical modeling like regression techniques, clustering, network analysis, simulation studies as well as explorative statistical strategies and visualization to detect patterns in (large) data sets.[4] In sociology, the inclusion

---

3   There is plenty of introductory literature and references to data science, data mining etc. See for example Larose and Larose (2014), Schutt and O'Neil (2014) or Cady (2017).
4   In difference to sociology and in contrast to their labeling policies, "data science" or "computational social science" are not sciences per se, but should be conceived of as fuzzy

of many new data sources has been a novelty (more than a decade ago): not only social media data and geo-data from smart phones, but also data from technical sensors (see Engel et al. 2022a, 2022b).

The risk of these trends and domains (not to say: buzz words) is that it could be an appealing option to replace sociology, and the methods and statistical tools embedded therein, by an overarching new disciplinary field, which does not bring in a substantial new definition of what social science and sociology is, but offers a "bag of statistical tools" devoted to the application of computer-generated and computer-intense analysis instead – as if the methods and statistical approaches mentioned had not been part of sociological social research for decades now.

Problematic aspects and negative consequences are several disembeddings that can be identified at this moment.

1. The disembedding of data analysis from sociological (and social) contexts, sociological paradigms and problematizations prepared by the "big data-world" by arguing that big data will make theory and theorizing needless (Anderson 2008; Mayer-Schönberger and Cukier 2013). Some influential scholars in the field of big data and computational social science suggest that data analysis technologies move to the center of the social sciences and are no longer interconnected with sociological theories and sociological ways of thinking (Lazer et al. 2009). The reduction of empirical research to the sophisticated analysis of massive amounts of computer-generated data as "facts" in the sense of a secure, ready-interpretable and sufficient empirical basis for science opens the way to a kind of "data-based positivism".[5] At the moment, the approach of analytical sociology, which is a methodological approach employing agent-based modeling, is the only sociological approach, which is substantially linked to computational social science (Hedström 2005; Jarvis et al. 2022). But this approach has its focus on modeling and its theoretical basis is still weak, which promotes the danger of disconnecting data analysis from sociological theorizing as well as

---

fields of reassembled existing competencies and contributions of other disciplines. Their substantial foundations are statistics, computer science (software programming, "data wrangling", "hacking skills") and "substance matter" sciences (Schutt and O'Neil 2014: 7). Evidently most "data scientists" are trained in the context or in disciplines as engineering, natural sciences, computer science and economics. See for a critical framing of "data science" as practice and profession Schutt and O'Neil (2014). Also, there is no established definition of "computational social science", its character is in flux. See for a discussion Engel et al. (2022) and Cioffi-Revilla (2022).

5   For the positivist notion of facts as a secure basis in science see Chalmers (1999: 17).

from substantial sociological research questions.[6] Evidently, sociology can improve and its analysis can advance by applying new data analysis technology. But this alone will result only in trivial findings. This argument was stated against a-theoretical big data analytics by John Savage with regard to the analysis of social inequality.

„The rise of big data caused growing anxiety about the status of social scientific knowledge, compounded by well-publicized examples of natural and information sciences deciding that they could now do social science using their computational skills without the need for much social scientific intervention. It is in the context of this challenge to their intellectual authority that the theme of inequality allowed social scientists to stage a brilliant riposte. By mobilizing a bold new stream of research that combined new forms of large-scale quantitative data analysis with qualitative theoretical insight, and yoked to a moral concern with the injustice of inequality, social scientists could seize on a big-picture story that had pretty much entirely evaded the attention of big data evangelists. The new social science of inequality trumpeted a bold, big, and commanding vision. By contrast, proponents of big data failed to deliver on the knowledge revolution that they promised, because their technically skilled findings were trivial, often defaulting to startling visual displays but with no defining narrative." (Savage 2021: 10–11)

2. If this data-based positivism regards data as secure basis, it will disregard the study of data production, data quality, new error sources and new forms of bias and move it to second rank. Although there is critical awareness for issues of data quality by established researchers in sociology (for survey data quality, see Blasius and Thiessen 2012), this kind of methodological reflexivity in these new domains remains an open question (Japec et al. 2015; see the discussion in Engel et al. 2022).[7]

3. The high level of interdisciplinary division of labor and the technical level of big data analytics opens the gates to hand over sociological analysis to outsiders, whose practice will be a kind of "ad hoc sociological research". The results are in many cases the reduction of large amounts of data to rather irrelevant patterns or word clusters (topic modeling) and graphs, because

---

6   See for a presentation of analytical sociology the contributions in Hedström and Bearman (ed.) (2009).
7   A series of empirical studies demonstrate how independent research groups proceed data analysis starting with the same dataset apply different data analysis workflows and that therefore data analysis is not determined by the given data. Instead the data analysis and outcome have shown a high variability of employed algorithms, data transformation and correction strategies as well as applied statistical techniques between the research groups resulting in different results and conclusions (Silberzahn et al. 2018; Botvinik-Never et al. 2020; Schweinsberg et al. 2021; Massey et al. 2022).

substantial research questions and scientific problematizations are missing. The scientific way of analysis will then be reversed: results are afterwards in search of a substantial research question and its sociological relevance. This could be called, now in an ironic sense, "data-driven social research".

4. In many cases, data sets are generated by companies or technical data infrastructure owned by private companies (retrieved and stored on the Internet by Internet companies), so that the conditions of data production, the kind of data quality and the resulting data analyses are not transparent and no analysis is reproducible, since they are proprietorial data and data access is regulated by entrepreneurial commercial policies (whose regulation may change over time). The problem here is that the data infrastructures with the most comprehensive data stocks are private property (Lazer et al. 2009; Diaz-Bone 2016). This is best illustrated by social media data, where the application programming interfaces (APIs, see Foster et al. 2017; Jünger 2022) are regulated by Internet companies and only limited access is available for non-commercial use. Missing practices and alliances for data sharing between public and private organizations and the question how to protect anonymity are obstacles for computational social science. Also, universities are not prepared to foster interdisciplinary collaboration and teaching for computational social science (see Lazer et al. 2009, 2020).

3.3 Data worlds

The different aspects discussed point to the reality of different "data worlds". The concept of data worlds grasps the plurality of statistical cultures, which employ statistics with different kinds of standards and epistemic values of what are criteria for "good statistical analyses" and what characterizes good statistical practices (Diaz-Bone and Horvath 2021; Vogel 2019). Table 1 presents five established data worlds as ideal types (in reality, they can show up not in their "pure" form).[8] The table emphasizes the differences by sketching different constellations of data infrastructures, common goods and quality criteria, to which coordination of actors in these worlds aim for.[9] In the last column, table 1 presents the most influential conventions as underlying (deeper) institutional logics, which pattern and structure coordination in the different data worlds in different ways.[10]

---

8   See Vogel (2019) for a more in-depth presentation of methodological worlds applied to surveys.
9   See for different classifications and sets of quality criteria in sciences Keller et al. (2017).
10 For more applications of convention theory to social research see Diaz-Bone et al. (2020) and Diaz-Bone and Horvath (2021).

Table 1: Data worlds

Data world	Actors, organizations	Main data infrastructure	Common good	Quality criteria	Conventions
Academic	Universities, scientific research institutions	Self-generated data or international survey programs	Discovering truth, contributing to enlightenment, providing objective data about society	Accuracy, objectivity, reliability, validity	Industrial, civic
Official statistics	National statistical institutes, statistical offices on regional or community level	Relying on administrative organizations and administrative data	Democratizing knowledge, enhance effectiveness of governance, providing objective data about society	Accuracy, completeness, long time series, compatibility to administration and governmental needs	Industrial, civic
Market research	Market research companies, enterprises	Entertains its own data generating survey infrastructure	Improving products and services for consumers, optimizing companies' resources allocation and profits	Efficiency for economic management and profit generation	Market, industrial
Big data	Big Internet companies	Employing Internet based and linked technological devices	[Normally no engagement for a common good]	Real-time and fine-grained prognoses; efficiency	Industrial, inspiration, market
Civic	NGOs, social movements, journalists	Combining publicly available data from other data worlds; self-generated data	Empowering civic agencies, countervail governmental or entrepreneurial representations of "social facts", fostering civic participation	Ready for use, interpretability, relevance for collective action, timeliness	Civic, industrial, domestic, network

203

Established data worlds are the academic data world, market research or the world of official statistics. More recent data worlds are the big data world and the civic data world. Data worlds link data production, interpretation and data-based decisions in different ways to values and pursue different forms of a common good. The plurality of data worlds is the result of data related to actors' coordination relying on a plurality of conventions (Diaz-Bone and Larquier 2022).

This plurality represents three additional external challenges for teaching statistics.

1. Data worlds are based on different conventions as institutional logics. The plurality of data worlds can no longer be conceived of as a set of data worlds under the hegemonic rule of scientific standards and epistemic values of the academic world only, but all data worlds consider themselves as capable and legitimate on their own. This is the reason, why data worlds question and criticize each other in many situations. The academic data world is criticized by the big data world for its limited amounts of data and its presumably "old-fashioned" technologies. Also, big data analytics in private companies (such as the big Internet companies AMAZON, Google, Meta) do not accept restrictions to safeguard individual privacy, data ethics and the academic world is regarded as naïve. The civic data world questions scientific claims of objectivity and criticizes the academic world and the world of official statistics for not delivering relevant data in time and fit for use – as it has been the case in the time of the COVID-19 pandemic or in the case of environmental issues (as the missing information about the radioactive contamination exposure in times of the Chernobyl nuclear disaster). Societies have never been so keen on data and numerical representations as nowadays, at the same time data have never been so contested in the public, too. The explanation for this seemingly paradoxical situation is the plurality of data worlds. Scientists are confronted by the mass media and public discourses with this kind of questioning universities' ways of reasoning (i.e. the academic world) about statistics and its ethics. The rise of the civic data world (including movements like open science, open access, citizen science) can be regarded as a counter-movement and a counter-position to the worlds of official statistics and big data. So, students are familiar with disputes between representatives of different data worlds in the public in general. But are students trained to develop competencies to judge when (and why) the conventions of the academic world are adequate or when other data worlds' criteria and conventions are to be applied (and why)?
2. The evaluation of statistical data in different data worlds requires knowledge about procedures, criteria and different common goods (which actors aim for) in different data infrastructures. One can speak of "data infra-

structure literacy" (Gray et al. 2018), which nowadays has to be regarded as a competence of sociologists to understand data production. Data quality literacy and data infrastructure literacy are prerequisites to evaluate the adequacy of statistical analyses and the informed interpretation of its results. It is only the academic world, which can build up its evaluations from the coherence of theory/methods packages. For the other worlds, it is their institutional context, social regulations and social needs as well as the (technical, juridical, practical) properties of data infrastructures, which coin quality and quality criteria. This kind of knowledge exceeds the usual concept of "context" relevant for data analysis, which is restricted to know where to find data sets and to recognize "metadata" only. As the convention theorist Alain Desrosières has stated, to quantify is to implement a convention and then to measure (Desrosières 2008: 10). The meaning of data, therefore, cannot be accessed without knowing the conventions at work in measurement. The quality of data cannot be evaluated without taking into account the relevant quality criteria. All in all, data analysis and statistics need to be embedded in a sociology of social research (such as convention theory), which studies the data infrastructures (and their conventions) to more fully understand what is "in the data" and what data represent.[11] Are students trained in this kind of data infrastructure literacy and how to apply it in "data wrangling" and statistical analysis?

3. The model of data worlds portrays them as ideal types. In reality, the plurality of data worlds brings in tensions and critiques, but also possibilities for alliances and compromises, which is the reason why real situations deviate from the ideal types. From a pragmatist perspective, critiques, justification, alliances, and compromises have to be exerted and proceeded by competent actors. Lazer et al. (2020) have already made such a proposal for an alliance between the academic world and the big data world, but without recognizing the contradicting conventions and criteria of these data worlds. Diaz-Bone and Horvath (2021) have pointed to the alliance between the civic world and the official statistics data world, which is made possible by shared common goods and conventions, to face the threads of big data. Sociology students are trained also in sociological theories (as different institutionalism, theories of social action and agency), which are an important cognitive resource to anticipate why some alliances and compromises between data worlds succeed, while others do not. Sociologists, therefore, are predestined and best prepared to be most professional in building new data infrastructures and in coordinating interdisciplinary and interprofessional teams in the field of data production and analysis. But are these new oppor-

---

11 For a perspective similar to convention theory see Beaulieu and Leonelli (2022).

tunities for sociology graduates anticipated in sociology programs and are students extensively trained in related skills how to manage data infrastructures and how to coordinate transdisciplinary teams?

4 Needs and perspectives

For sociology, the core question arises: how to guide and train students (in the time available in BA and MA programs) so that they will be able to cope with the complexities and the plurality of data worlds to (1) apply coherent ways of combining theories, explanatory logics and statistical tools as well as to (2) engage in designing and managing data infrastructures with interdisciplinary teams? To find answers and solutions to this question requires to recognize sociology as a unique discipline and profession in changing and complex professional contexts (as Andrew Abbott 1988 has analyzed).

From a disciplinary perspective, some general conclusions for statistics teaching can be drawn out of the foregoing considerations. Thereby, the discussed challenges and complexities will be addressed. It is necessary to remember some classical methodological positions in sociology and to combine them with a proposal for a more sociological perspective and response.

4.1 "Hands-on-teaching" cannot substitute foundational training first

Data scientists promote so-called "hands-on-teaching", combining statistics education, software application education and the analysis of social research questions as a way to learn about statistics in practical lessons. Data scientists normally have a background in engineering, computer science, economics or natural sciences and they have passed several lectures in their first BA semesters in mathematics and statistics before specializing in programming and data science. Sociologists do not, which is the reason why hands-on seminars will have almost no learning outcomes beyond some examples and some given software codes. It will not work out to understand the many and complex statistical concepts by examples only and by practically using computers without understanding what computers' output is exactly about. Hubert Blalock has offered the classical and still influential position on this.

> "I am deeply concerned about an increasing communication gap as a result, with today's graduate students being far less close to the cutting edges of methodological developments than was the case several decades ago. There is simply too much technical material to be absorbed during the very limited time that our average graduate student devotes to statistics and methods courses. And virtually none of them take mathematical modeling courses that presuppose strong backgrounds in probability

theory, differential equations, or linear or matrix algebra. Both of these facts, I believe, contribute to a tendency to substitute training in techniques and the uncritical use of computer programs rather than more in-depth knowledge that places a higher premium on the understanding of underlying assumptions and their implications for theoretical interpretations of research findings." (Blalock 1989: 448/449)

Also, project-related learning will fail in the same way, when students do not already dispose of knowledge in statistics, data handling and sociology. Again, a classical position has been coined by Blalock.

> "One common argument is that students should learn their methodology by attaching themselves to an on-going project, learning, in effect, as apprentices. Fine. But I do not believe this is enough. Many technical arguments need to be developed systematically in a well-organized set of lectures or readings. Learning 'causal inferences' by doing would, in my opinion, be a highly wasteful and nonsystematic effort." (Blalock 1969: 5)

To be clear, practical applications and project-based learning can be highly effective, but they have to be embedded in a smart way in sociology programs' curricula. First, foundational introduction into statistics, sociology and data analysis will be needed. Later on, students can exercise the integration of these competences in project-based learning, where the objective is the integration of different skills and the experience of the work flow (to pursue a research interest). Applying the basics is not learning the basics, but to combine them with illustrations and to enlarge their comprehension.[12]

4.2 Econometrics is not an example for sociology

Many statistics courses in sociology are implicitly or explicitly informed by econometrics and restrict themselves to regression analysis (and general linear models). There are many quarrels to address. First, economics and the related regression logics avoid a good coverage of statistical approaches that are fit to deal with categorical data, which are more common to sociological data sets than metrical data. Second, regression analysis emphasizes linear relations between variables and suggests causal models that are too simple (with one dependent variable only). Third, visual analysis of data is weakly developed in regression analysis (restricted mainly to the analysis of residuals). And fourth, linear regression models have too many and too strong requirements to fulfill

---

12 This is supported nowadays by modern data analysis software as R, where the workflow can be organized by file formats as "R notebooks" (in RStudio) to integrate text, code, and output (Wickham and Grolemund 2017).

for real data sets – and the analysis of violations and the training of correction consume too much time in lessons. Any statistical training relevant for sociology has to include an introduction to multivariate analysis of categorical data. Multiple correspondence analysis (Blasius 2001) and logistic regression nowadays are a must to learn for sociologists in BA programs.[13] In addition, this multiplicity of approaches opens more options to link statistics to different explanatory logics, such as methodological holism, and not only to methodological individualism (which is linked to regression analysis). This statistical approach is also confined to hypothesis testing and to modeling one dependent variable only. In contrast, multiple correspondence analysis (MCA) is exploratory and visual in character. It has been Jean-Paul Benzécri and John Tukey who emphasized the impact and power of explanatory data analysis and visualization to detect sociologically relevant patterns and underlying (latent) dimensions in the data (Tukey 1977; Benzécri 1992) – a classical position (almost forgotten by data science and computational social science), which has been foundational for sociological approaches to teach statistics. In difference to the sketched a-theoretical positions of big data and computational social sciences, the different steps of MCA are based on theoretical conceptions of social space and social fields, whose empirical dimensions are identifiable (as factors) by MCA, but are in need to engage theoretical interpretation. The steps of MCA (from selecting indicators for the factors to interpretation of results) are theory-driven, not data-driven (Bourdieu 1984; Blasius et al. 2019).

4.3 Embedding statistical teaching

The need to interpret data and the results of statistical procedures, thereby applying the knowledge of the properties of data infrastructures, data production practices and conventions of data worlds has already been underlined. But it is important to do this by offering to students different learning opportunities how to link statistical analysis with sociological theory, too. Without sociological interpretation, statistics cannot be meaningful at all and students will not fully grasp the theory/methods packages. Moran has made explicit the fundamental role of theorizing when teaching statistics.

> "From a sociological perspective, it is simply impossible to conduct a-theoretical quantitative analysis. Most methodological and statistical decisions do not have unique answers, and more often than not 'objective'

---

13   More advanced statistical techniques should be offered as options in sociological MA programs. Unfortunately, most US-American (e.g. Agresti 2013) or German (e.g. Andreß et al. 1997) text books for categorical data analysis do not entail multiple correspondence analysis. But see Greenacre and Blasius (ed.) (2006) or Friendly and Meyer (2016).

technical criteria are of limited help in guiding a researcher through the maze of decisions that need to be made. Theory, then, is the ultimate arbiter. Everything from measurement choices, to model specifications, to the handling of missing values is better addressed through sociological theory than statistical fiat, and the sociologist teaching statistics should emphasize the former and downplay the latter. As most of us know quite well, theory cannot be made out of the results of statistical analysis but can only be invented by intuition, reasoning, and creative thinking. While statistics courses in sociology (at least at the graduate level) hopefully emphasize the fundamental role of sociological theory in quantitative analysis, more of a concern is the disappearance from textbooks and courses of the equally essential role of statistical theory in producing and interpreting social scientific research." (Moran 2005: 264–265)

As Jörg Blasius and Victor Thiessen (2021) have demonstrated in their recent book, statistics text books can closely link data analysis and sociological interpretation. This way, it is avoided to downsize social contexts to simple examples and "real data" only. Categorical data analysis mobilized by a sociological research question, informed by sociological theory and knowledge on (data generating) social contexts can be applied to real data sets in specific seminars (like project related seminars), so that students are effectively introduced to theory/methods packages. Sociology's complexities as challenges then can be transformed into sociology's pluralities of paradigms and of theory/methods packages as a scientific richness, which makes sociology still relevant and a provider of theory, methods and analyses for other disciplines but also for different institutions, actors, groups and other fields in society.

References

Abbott, Andrew (1988): The system of professions. An essay on the division of expert labor. Chicago: Chicago University Press.
Agresti, Alain (2013): Categorical data analysis. 3rd edition. New York: Wiley.
Alvarez, Michael R. (ed.) (2016): Computational social science. Discovery and prediction. Cambridge: Cambridge University Press.
Anderson, Chris (2008): The end of theory. The data deluge makes the scientific method obsolete. In: Wired, 23 June 2008.
Andreß, Hans-Jürgen/Hagenaars, Jacques A./Kühnel, Steffen (1997): Analyse von Tabellen und kategorialen Daten. Berlin: Springer.
Baur, Nina/Blasius, Jörg (ed.) (2022): Handbuch der empirischen Sozialforschung. 3rd edition. Wiesbaden: Springer VS.
Beaulieu, Anne/Leonelli, Sabina (2022). Data and society. Los Angeles: Sage
Ben-Zvi, Dani/Makar, Katie (ed.) (2016): The teaching and learning of statistics. International perspectives. Cham: Springer.

Ben-Zvi, Dani/Makar, Katie/Garfield, Joan (ed.) (2018): International handbook of research in statistics education. Cham: Springer.
Benzécri, Jean-Paul (1992): Correspondence analysis handbook. New York: Marcel Dekker.
Blalock, Hubert M. (1969): On graduate methodology training. In: American Sociologist 4, 1, pp. 5-6.
Blalock, Hubert M. (1989): The real and unrealized contributions of quantitative sociology. In: American Sociological Review 54, 3, pp. 447-460.
Blasius, Jörg (2001): Korrespondenzanalyse. München: Oldenbourg.
Blasius, Jörg/Lebaron, Frédéric/Le Roux, Brigitte/Schmitz, Andres (eds.)(2019): Empirical investigations of social space. Cham: Springer.
Blasius, Jörg/Thiessen, Victor (2012): Assessing the quality of survey data. Los Angeles: Sage.
Blasius, Jörg/Thiessen, Victor (2021): Argumentieren mit Statistik. Eine Einführung für das sozialwissenschaftliche Studium. Opladen: Barbara Budrich.
Botvinik-Never, Rotem et al. (2020): Variability in the analysis of a single neuroimaging dataset by many teams. In: Nature 582, pp. 84- 88.
Bourdieu, Pierre (1984): Distinction. Cambridge: Harvard University Press.
Cady, Field (2017): The data science handbook. New York: Wiley.
Chalmers, Alan (1999): What is this thing called science? 3rd edition. St. Lucia: University of Queensland Press.
Cioffi-Revilla, Claudio (2017): Introduction to computational social science. Principles and applications. 2nd edition. Cham: Springer.
Cioffi-Revilla, Claudio (2022): The scope of computational social science. In: Engel, Uwe/Quan-Haase, Anabel/Liu, Sunny Xun/Lyberg, Lars (ed.): Handbook of computational social science. Vol. 1: Theory, case studies and ethics. Abingdon: Routledge, pp. 17-32.
Clarke, Adele/Friese, Carrie/Washburn, Rachel S. (2018): Situational analysis. Grounded theory after the interpretive turn. 2nd edition. Sage: Los Angeles.
Desrosières, Alain (2008): Pour une sociologie historique de la quantification. L'argument statistique I. Paris: Mines ParisTech.
Diaz-Bone, Rainer (2016): Convention theory, classification and quantification. In: Historical Social Research 41, 2, pp. 48-71.
Diaz-Bone, Rainer (2018): Die "Economie des conventions". Grundlagen und Entwicklungen der neuen französischen Wirtschaftssoziologie. 2nd edition. Wiesbaden: Springer VS.
Diaz-Bone, Rainer/Horvath, Kenneth/Cappel, Valeska (2020): Social research in times of big data. The challenges of new data worlds and the need for a sociology of social research. In: Historical Social Research 45, 3, pp. 314-341.
Diaz-Bone, Rainer/Horvath, Kenneth (2021): Official statistics, big data and civil society. Introducing the approach of "economics of convention" for understanding the rise of new data worlds and their implications. In: Statistical Journal of the International Organization of Official Statistics 37, 1, pp. 219-228.

Diaz-Bone, Rainer/Larquier, Guillemette de (2022): Conventions. Meanings and applications of a core concept in economics and sociology of conventions. In: Diaz-Bone, Rainer/Larquier, Guillemette de (ed.): Handbook of economics and sociology of conventions. Cham: Springer, pp. 1–27.

Diaz-Bone, Rainer/Weischer, Christoph (ed.) (2015): Methoden-Lexikon für die Sozialwissenschaften. Wiesbaden: Springer VS.

Edelmann, Achim/Wolff, Tom/Montagne, Danielle/Bail, Christopher A. (2020): Computational social science and sociology. In: Annual Review of Sociology 46, pp. 61–81.

Engel, Uwe/Quan-Haase, Anabel/Liu, Sunny Xun/Lyberg, Lars (2022): Introduction to the Handbook of computational social science. In: Engel, Uwe/Quan-Haase, Anabel/Liu, Sunny Xun/Lyberg, Lars (ed.), Handbook of computational social science. Vol. 1: Theory, case studies and ethics. Abingdon: Routledge, pp. 1–14.

Engel, Uwe/Quan-Haase, Anabel/Liu, Sunny Xun/Lyberg, Lars (ed.) (2022a): Handbook of computational social science. Vol. 1: Theory, case studies and ethics. Abingdon: Routledge.

Engel, Uwe/Quan-Haase, Anabel/Liu, Sunny Xun/Lyberg, Lars (ed.) (2022b): Handbook of computational social science. Vol. 2: Data science, statistical modelling, and machine learning methods. Abingdon: Routledge.

Foster, Ian/Ghani, Rayid/Jarmin, Ron S./Kreuter, Frauke/Lane, Julia (2017): Big data and social science. A practical guide to methods and tools. Boca Raton: CRC Press.

Friendly, Michael/Meyer, David (2016): Discrete data analysis with R. Visualization and modeling techniques for categorical and count data. Boca Raton: CRC Press.

Gray, Jonathan/Gerlitz, Carolin/Bounegru, Lilian (2018): Data infrastructure literacy. In: Big Data and Society 5, 2, pp. 1–13.

Greenacre, Michael/Blasius, Jörg (ed.) (2006): Multiple correspondence analysis and related methods. London: Chapman and Hall.

Hedström, Peter (2005): Dissecting the social. On the principles of analytical sociology. Cambridge: Cambridge University Press.

Hedström, Peter/Bearman, Peter (ed.) (2009): The Oxford handbook of analytical sociology. Oxford: Oxford University Press.

Japec, Lilli/Kreuter, Frauke/Berg, Marcus/Biemer, Paul/Decker, Paul/Lampe, Cliff/ Lane, Julia/ O'Neil, Cathy/Usher, Abe (2015): Big data in survey research. In: Public Opinion Quarterly 79, 4, pp. 839–880.

Jarvis, Benjamin F./Keuschnigg, Marc/Hedström, Peter (2022): Analytical sociology amidst a computational social science revolution. In: Engel, Uwe/Quan-Haase, Anabel/Liu, Sunny Xun/Lyberg, Lars (ed.): Handbook of computational social science. Vol. 1: Theory, case studies and ethics. Abingdon: Routledge, pp. 33–52.

Jünger, Jakob (2022): A brief history of APIs. Limitations and opportunities for online research. In: Engel, Uwe/Quan-Haase, Anabel/Liu, Sunny Xun/Lyberg, Lars (ed.): Handbook of computational social science. Vol. 2: Data science, statistical modelling, and machine learning methods. Abingdon: Routledge, pp. 17–32.

Kelleher, John D./Tierney, Brendan (2018): Data science. Cambridge: MIT Press.

Keller, Sallie/Korkmaz, Gizem/Orr, Mark/Schroeder, Aaron/Shipp, Stephanie (2017): The evolution of data quality. In: Annual Review of Statistics and Its Application 4, pp. 85–108.

Kuhn, Thomas S. (1962): The structure of scientific revolutions. Chicago: Chicago University Press.

Larose, Daniel T./Larose, Chantal D. (2014): Discovering knowledge in data. An introduction to data mining. 2nd edition. New York: Wiley.

Lazer, David/Pentland, Alex/Adamic, Lada/Aral, Sinan/Barabási, Albert-László/Brewer, Devon/Christakis, Nicholas/Contractor, Noshir/Fowler, James/Gutmann, Myron/Jebara, Tony/King, Gary/Macy, Michael/Roy, Deb/Van Alstyne, Marshall (2009): Computational social science. In: Science 323, pp. 721–723.

Lazer, David/Pentland, Alex/Watts, Duncan J./Aral, Sinan/Athey, Susan/Contractor, Noshir/Freelon, Deen/Gonzalez-Bailon, Sandra/King, Gary/Margetts, Helen/Nelson, Alondra J./Salganik, Matthew/Strohmaier, Markus/Vespignani, Alessandro/Wagner, Claudia (2020): Computational social science: Obstacles and opportunities. In: Science 369, pp. 1060–1062.

Massey, Douglas et al. (2022): Observing many researchers using the same data and hypothesis reveals a hidden universe of uncertainty. In: Proceedings of the National Academy of Sciences of the United States of America (PNAS) 119, 44, pp. 1–8.

Mayer-Schönberger, Viktor/Cukier, Kenneth (2013): Big data. A revolution that will transform how we live, work, and think. London: John Murray Publishers.

Moran, Timothy P. (2005): The sociology of teaching graduate statistics. In: Teaching Sociology 33, 3, pp. 263–284.

Savage, John (2021): The return of inequality. Social change and the weight of the past. Cambridge: Harvard University Press.

Schmitz, Andreas/Schmidt-Wellenburg, Christian/Witte, Daniel/Keil, Maria (2019): In welcher Gesellschaft forschen wir eigentlich? Struktur und Dynamik des Feldes der deutschen Soziologie. In: Zeitschrift für Theoretische Soziologie 2, pp. 245–276.

Schutt, Rachel/O'Neil, Cathy (2014): Doing data science. Bejing: O'Reilly.

Schweinsberg, Martin et al. (2021): Same data, different conclusions. Radical dispersion in empirical results when independent analysts operationalize and test the same hypothesis. In: Organization Behavior and Human Decision Processes 165, pp. 228–249.

Silberzahn, Raphael et al. (2018): Many analysts, one data set. Making transparent how variations in analytic choices affect results. In: Advances in Methods and Practices in Psychological Science 1, 3, pp. 337–356.

Tukey, John (1977): Exploratory data analysis. Reading: Addison-Wesley.

Vogel, Raphael (2019): Survey-Welten. Eine empirische Perspektive auf Qualitätskonventionen und Praxisformen der Umfrageforschung. Wiesbaden: Springer VS.

Wickham, Hadley/Grolemund, Garrett (2017): R for data science. Bejing: O'Reilly.

Effects of Visual Priming on Rating Scale Usage

Pieter C. Schoonees, Patrick J.F. Groenen, Michel van de Velden,
Hester van Herk

Abstract

Rating scales are much used in survey research. Often, it is assumed that the scores obtained through rating scales can be compared within and between respondents when studies are in one country. In addition, it is assumed that they can be treated as a numerical scale. In this paper, we study the anchoring effect of a visual stimulus on rating scale usage. To do so, we set up a randomized experiment where the experimental group was primed by asking to rate the filling of a cylinder that was presented visually. For a five-point rating scale, we find the effect that primed respondents use Category 1 less and Categories 3 and 4 more, and no effect on Categories 2 and 5.

1. Introduction

The analysis and in particular the quality of survey data is an important topic in social science research to which Jörg Blasius has made many contributions over the years (see, for example, Blasius and Thiessen, 2012). Here, too, we study the possibilities for improving quality of respondents' ratings. In particular, we investigate one of the key assumptions of most analyses of surveys using rating scales, that is, the answers on the items can be compared within and between respondents. Often these assumptions are not tested when comparing respondents in the same country. One source of differences in rating scale usage by respondents is due to various response styles, such as extreme scoring, acquiescence, and midpoint scoring (Van Herk et al., 2004). Several solutions have been proposed for dealing with response styles. A simple one is ipsatizing (Cunningham et al., 1977; Rudnev, 2021).

Technical solutions to response style problems include searching for a priori unknown response style patterns through the latent class bilinear multinomial logit model (Van Rosmalen et al., 2010) and the constrained dual scaling model by Schoonees et al. (2015). In this paper, we study a different approach, that of priming the respondents by a visual task that aims at anchoring the rating scale to a clear visual stimulus. The aim is to use priming for calibrating respondents to the same scale usage. To do so, we investigate through an experiment whether there is an effect of anchoring, here, priming the respondents with a task where they need to rate the amount of filling of a cylinder by indicating how much of the cylinder was filled with red (see Figure 1 for an example). After this task, the questionnaire continued with questions using 5-point rating scales. In the remainder of this paper, we discuss first how the data were gathered, the experiment was set up, and how the priming through a visual stimulus (the amount of filling of a cylinder) was done. Then, the consistency of the answers in the experimental condition (rating versus percentage) is explored. After that, we perform two tests to compare the distributions over the average rating scale categories between the treatment and control groups. Subsequently, the difference in distributions is explored at the item level to see, amongst others, whether there is a diminishing effect over time. Finally, a correspondence analysis is used to study additional item differential effects. We end with a discussion and conclusion.

Figure 1: Example of visual stimulus of a cylinder with the bottom 30% red-filled used in one of the treatment items. The respondent was asked to estimate the amount of red in the cylinder on a five-point rating scale.

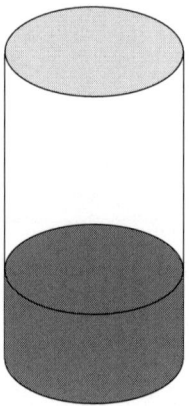

2. Materials and Methods

The data were collected in 2013 by GfK Panel Services Benelux B.V., a professional marketing research company. The total number of respondents in this research is 1621. They were 58.5% females, 13.6% 18–29 years, 17.2% 30–39 years, 20.4% 40–49 years, 27.1% 50–64 years, and 21.7% 65 years or older, meaning that they are a bit older than the general Dutch population. Of these respondents, roughly half (786) were presented with the cylinder questions (the treatment group). The other 835 respondents did not receive these questions (control group). Individuals were randomly allocated to one of the groups, with no statistically significant differences in gender, age group, or education level.

In the questionnaire, the first five survey items concerned the respondent's economic outlook for the Dutch economy, which was measured on a three-point scale with categories "better", "worse," or "unchanged". These items were answered by all respondents. For the treatment group, this was followed by the 10 cylinder questions (see Figure 1 for an example), which consisted of two blocks of five questions each. In the first block, respondents were to assess the amount of red on each cylinder on a 5-point scale ranging from "almost no red" (1) to "almost entirely red" (5), with the middle categories not labelled. The true percentages of the amount of red in the cylinders was 10%, 30%, 50%, 70%, and 90% corresponding naturally to a five point rating scale. In the second block of five questions, the respondents were asked to give percentage (0–100) to indicate the redness of the cylinder. The questions within each of these two blocks were presented in random order, and each cylinder was displayed for one second.

The control group did not receive the 10 cylinder items. After the cylinders, both groups were asked to assess 16 items from the marketing literature on innovativeness (Steenkamp et al., 1999) and materialism (Richins and Dawson, 1992). These multi-item scales were measured on a 5-point Likert scale (ranging from "totally agree" to "totally disagree"). Finally, all respondents were asked to answer 4 questions on sustainable behavior by ticking one of the options "never", "sometimes", "often", "very often" or "almost always".

In the next sections, we mainly focus on the 16 items of the innovativeness and materialism scales as these items followed directly on the cylinder items in the treatment group and used the same five point rating scale.

3. Results

First, we study the treatment group in their consistency of the rating scale usage and their estimated percentages of red. Then we focus whether or not there is a difference in the distribution in rating scale usage on the 16 items between treatment and control groups. After that, we study the differences on an item level.

Figure 2: Estimated ratings plotted against estimated percentages for the cylinder questions (jittered). The red-dashed boxes shows where correctly transformed percentages should lie.

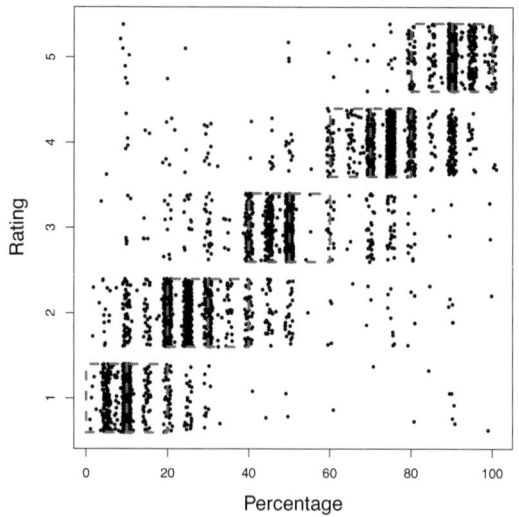

3.1 Answer Consistency within the Treatment Group

To study the consistency of answering the cylinder items, we compare the estimated percentage with the rating scale value on the 10 cylinder items. Recall that in random order, each true percentage of red filling was not only rated by the respondent on a five point rating scale but they also gave estimated percentage of the amount of red. Figure 2 shows a scatterplot of the estimated percentages (horizontal) against the estimated rating vertically. To ensure some spread of individual observations in the plot, the ratings are slightly jittered. The superimposed red boxes indicate the estimated percentages that should be mapped to the corresponding ratings corresponding to the correct filling. Hence, all observations outside the red boxes represent errors. There could be several sources of such errors. The respondent could misinterpret the direction of the rating scale of the percentage scale, or could make a typing error. Additionally, the visual stimulus might not be easy to assess for a respondent. Finally, other sources of errors usual in surveys (participant fatigue, response styles, etc.) could also have an effect.

There are different ways to assess whether these really are mistakes. First, consider errors made in estimating the percentages, thereby only considering the horizontal axis in Figure 2. Per true cylinder filling, the error rate is one minus the proportion of observations that provide the estimated percentage within ten percent of the true percentage of red. These error rates for the five cylinders are 7%, 22%, 18%, 21% and 4% for cylinder fillings of 10%, 30%, 50%, 70%, and 90%, respectively. This result indicates that the middle cylinders with true filling of 30%, 50%, and 70% are the hardest to correctly associate with the admissible range of percentages, as can be expected. Next, consider the errors of the ratings. Here, it is interesting to look at the errors conditional on the estimated percentages. To do so, we use again the intervals of estimated percentages that were used for the red boxes in Figure 2. Table 1 shows the proportion of responses in the successive intervals [0, 20], (20, 40], ..., (80, 100].

Table 1: Proportion of responses in the intervals [0, 20], (20, 40], . . . , (80, 100].

Interval	Rating				
	1	2	3	4	5
[0, 20]	0.65	0.31	0.02	0.01	0.01
(20, 40]	0.06	0.69	0.23	0.02	0.00
(40, 60]	0.01	0.09	0.81	0.09	0.01
(60, 80]	0.00	0.02	0.09	0.81	0.07
(80, 100]	0.01	0.01	0.01	0.18	0.79

The number of respondents who were able to associate the correct percentage interval with the respective ratings range from 65% to 81%. These numbers are not particularly high. However, given that the cylinders are displayed for one second only and that it is not possible to view the cylinder a second time, it may be expected that each respondent will make one mistake. In that case, the hitrate would be 80%. For rating Categories 1 and 2, the observed hitrates are lower (65% and 69% respectively).

Furthermore, the number of errors made by each respondent can be studied. Define an error as misspecifying a rating by more than one rating category or as misjudging a percentage by more than 15%. Errors were made by 134 out of 786 respondents (or 17%). By this definition, a total of 184 mistakes were made. There were 101, 19, 11 and 3 respondents who made 1, 2, 3 or 4 mistakes respectively. Despite these errors, the subsequent results are based on all 786 respondents in the treatment group. The main reason being that the large majority of respondents (95.8%) made no errors or only one error.

3.2 Treatment Differences on Average Rating Scale Distributions

In this subsection, we study the causal effect of priming the respondents with the cylinder items. In particular, we look at the overall distribution of rating scale usage over the 16 items combined. As a first test, we investigate the pre-treatment distribution over the rating scale categories.

For this purpose, we consider the five survey items on the economic outlook for the Dutch economy as they appeared in the survey before the experimental condition. As respondents were randomly assigned to the treatment or control groups, the expectation is that there should be no significant differences in distribution over the three-point rating scale items. Figure 3 shows a line per rating scale category over the five items with the proportion usage split by treatment (solid line) and control group (dashed line). As the dashed and solid lines almost overlap, we can conclude that the random assignment has worked well and that there were no pre-treatment differences between the two groups.

We now switch to the 16 post-treatment items and compare the average rating scale usage. As a first step, we do a two-sample Kolmogorov-Smirnov (KS) test to test the null hypothesis that the two distributions of usage of the rating scale categories are equivalent. Comparing the post-treatment scale usage for the 16 items, the null hypothesis is rejected ($D = 0.0617$, asymptotic $p < 0.0001$).

A more thorough test on differences between the two distributions that is less sensitive for tied data than the KS-test can be obtained through a permutation test. In this test, a statistic derived from the two distributions is compared with the distribution of this statistic under random permutation of the grouping (that is, treatment or control). The test randomly permutes the groupings (i.e. treatment of control) a large number of times (10,000 in this case) and compares the observed difference in category usage between the treatment and control groups with the permutation distribution of the same statistic. If the observed difference of the test statistic falls in the tails of the permutation distribution, it can be concluded that the null hypothesis of equal distributions is not plausible and therefore rejected. The test can be investigated per rating, but also across all ratings by using the sum of squared differences.

Figure 3: The proportion of rating categories used for the five items delivered before the cylinder questions, separately for the treatment and control groups.

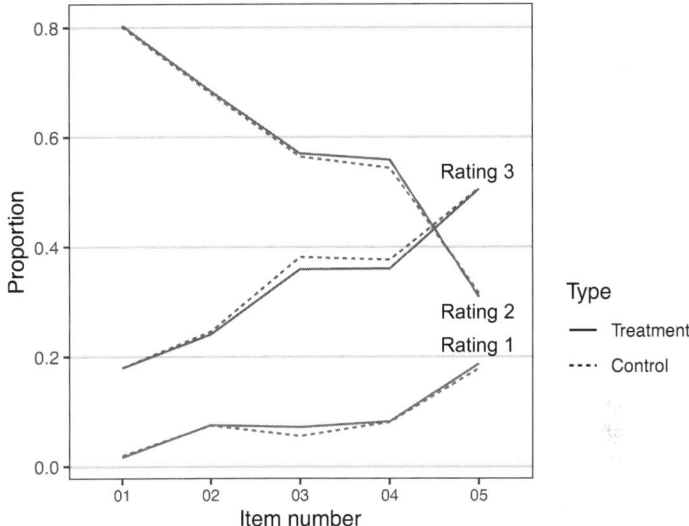

Figure 4: Global permutation distribution for sum of squared differences in relative frequencies of rating scale use, using 10,000 permutations. The observed value of the test statistic is shown by the red vertical line.

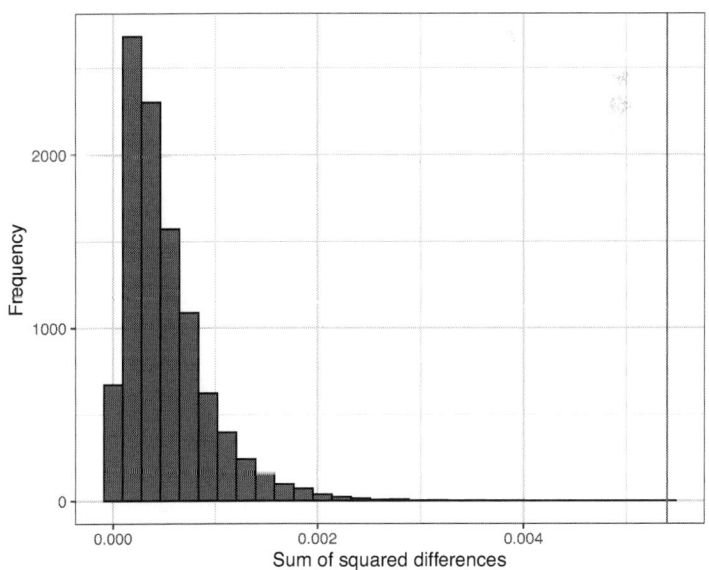

Here, we apply the permutation test on a comparison on average rating scale distribution over the items. As a test statistic, we simply use the sum of squared differences of relative frequencies of the rating scales between the treatment and control groups. This permutation distribution for 10,000 random permutations is shown in Figure 4 where the (unpermuted) observed test statistic is indicated by the red line. As all 10,000 test-statistics obtained by the random permutation of the two groups are smaller than the observed test statistic, we can reject the null hypothesis of equal distributions with $p < 0.0001$.

In Figure 5, the two observed distributions are shown. It is apparent that administering the cylinder items changes the subsequent rating scale usage of the respondents. Specifically, the treatment group uses fewer 1's (19.6% versus 27.5%), more 3's (32.5% versus 28.7%) and more 4's (19.6% versus 16.3%). In conclusion, there is a significant difference in rating scale use after the cylinder questions have been administered. The cylinder items seem to make respondents more aware of the spread of the rating scale. Combined with the evidence from Section 3.1 it can be argued that the anchoring through the cylinder items seems to make respondents pay more attention to the meaning of rating scale categories within and across respondents.

Figure 5: Rating scale usage for the treatment and control groups for the 16 survey items following the cylinder intervention.

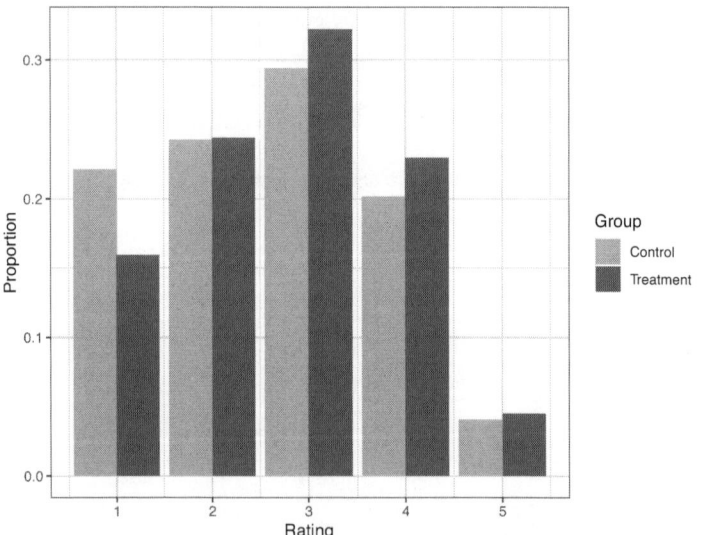

3.3 Treatment Differences on per Item Rating Scale Distributions

Having established in the previous section that there is an *average* effect of the cylinder items on the distribution over the categories of the rating scale, we now turn to item specific effects. Table 2 and Figure 6 give the relative frequencies (proportions) per rating and item for the treatment and control groups. In the previous section, we concluded that the differences mainly lie in the usage of Categories 1, 3, and 4. In particular, Figure 6 shows that for Category 1, there is a much smaller proportion for almost all items, with the strongest effects for the first 14 items after the treatment. In addition, Categories 3 and 4 are used more often in the treatment group, again with the strongest effects for the first items after the treatment. In conclusion, the anchoring effect of the cylinder items seems strongest for the first (say 10–15) items, after which the effect becomes hard to detect. Moreover, Category 1 is used less often in the treatment group and Categories 3 and 4 more.

3.4 Differential Item Effects

In addition to the average effect of the priming on rating scale usage, there can be a differential effect for the items. To explore this, we perform a correspondence analysis (CA) on the table of items by rating categories for the two groups. That is a frequency table with the rating categories for the two groups in the columns and the items in the rows. It is well known that the CA of this matrix removes the main and interaction effects of category usage and being in the treatment or control group (see Van der Heijden et al., 1989). The CA will therefore only show the remaining effects between items and rating scale category of treatment and control groups.

Figure 7 shows the results of the CA. First, we observe that the first dimension explains almost 80 percent of the inertia and is about five times stronger than Dimension 2. Although the points of categories for treatment and control are close, those of the treatment are slightly, but consistently, to the left of those of the points corresponding to the categories for the control. Apart from the main and interaction effects, items V19 and V20 have more scores on Category 1 and fewer scores on Categories 4 and 5. This effect is slightly stronger for the Treatment group. For V30 the opposite is true: it has more scores in Categories 5 and 4 and fewer in Category 1, and the effect is slightly stronger for the Control group than the Treatment group.

Table 2: Proportion of rating scale use per item for the treatment (left) and control groups (right), for 16 items administered after the cylinder question. The marginal proportions are shown in the final row.

	Rating Treatment						Rating Control				
Item	1	2	3	4	5	Item	1	2	3	4	5
Innovation						Innovation					
V16	0.22	0.29	0.37	0.12	0.02	V16	0.36	0.27	0.28	0.08	0.01
V23	0.12	0.24	0.40	0.20	0.03	V23	0.16	0.27	0.39	0.16	0.03
V24	0.13	0.27	0.41	0.17	0.02	V24	0.17	0.27	0.37	0.17	0.02
V25	0.10	0.19	0.39	0.28	0.03	V25	0.14	0.22	0.36	0.26	0.03
Materialism						Materialism					
V17	0.20	0.30	0.29	0.19	0.03	V17	0.32	0.29	0.25	0.12	0.02
V18	0.24	0.32	0.30	0.12	0.02	V18	0.35	0.34	0.22	0.08	0.02
V19	0.36	0.35	0.22	0.06	0.01	V19	0.48	0.31	0.17	0.03	0.01
V20	0.45	0.32	0.18	0.03	0.01	V20	0.54	0.29	0.14	0.03	0.01
V21	0.07	0.18	0.33	0.39	0.03	V21	0.11	0.15	0.34	0.36	0.04
V22	0.07	0.15	0.36	0.39	0.03	V22	0.11	0.14	0.37	0.35	0.04
V26	0.10	0.28	0.38	0.23	0.02	V26	0.17	0.29	0.37	0.16	0.02
V27	0.09	0.18	0.31	0.30	0.13	V27	0.11	0.18	0.32	0.29	0.10
V28	0.08	0.21	0.34	0.31	0.06	V28	0.11	0.25	0.31	0.29	0.05
V29	0.11	0.29	0.28	0.27	0.04	V29	0.14	0.28	0.28	0.27	0.04
V30	0.01	0.03	0.26	0.50	0.20	V30	0.02	0.04	0.25	0.49	0.20
V31	0.21	0.30	0.34	0.12	0.04	V31	0.24	0.31	0.33	0.09	0.03
Total	0.16	0.24	0.32	0.23	0.04	Total	0.22	0.24	0.29	0.20	0.04

Figure 6: The proportion of rating categories used for the 16 items delivered after the cylinder questions, separately for the treatment and control groups. Each panel shows a different rating category

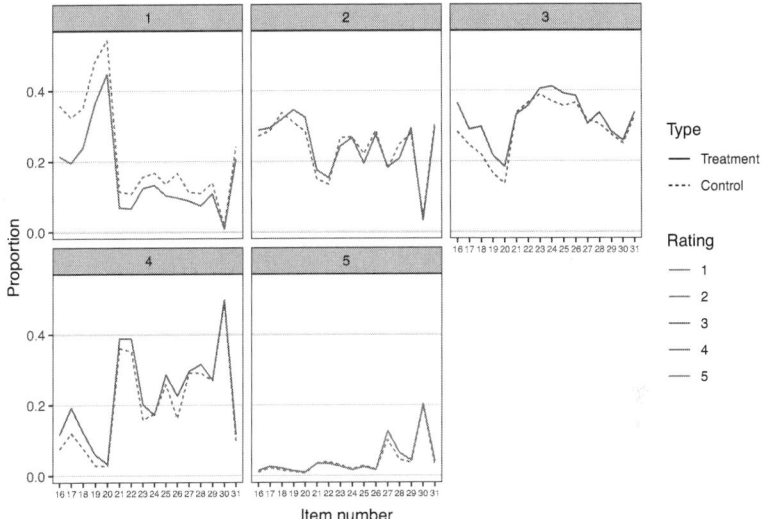

Figure 7: Correspondence analysis on the rating scale usage of Items 16 to 31. The frequencies for the treatment (t) and control (c) groups are calculated separately. These are then concatenated row-wise, and subjected to correspondence analysis.

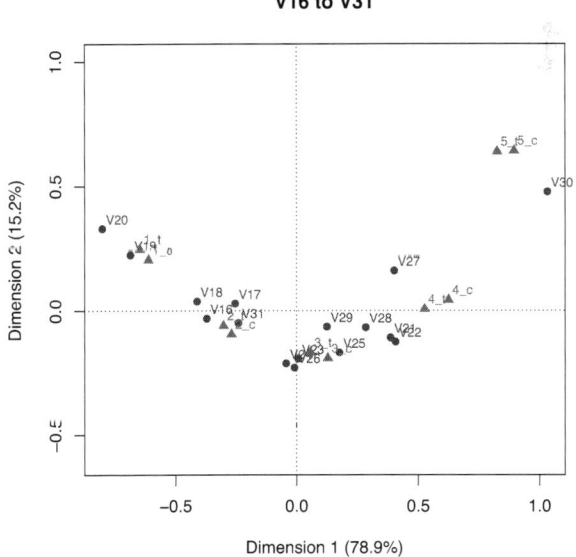

223

4. Conclusions and Discussion

It is evident that the simple cylinder questions has a priming effect on the respondents which greatly affects their rating scale usage. Our experiment shows that respondents answering on a five point rating scale in the primed group use Category 1 less and Categories 3 and 4 more than those who were not primed. This suggests that the treatment reduces extreme responding, a main response style, and may motivate respondents to pay closer attention to items.

The section on consistent rating scale usage provides evidence that about 17% of the respondents make errors in their ratings, but only 4% make more than one error. It is possible that the visual cues by the cylinders are misinterpreted by these respondents. In this analysis, we have used all observations and not made a selection of the participants based on these inconsistencies in the treatment group. It is our expectation that if those participants would have been removed, the effect of the visual priming could be stronger.

We conclude that priming respondents by a visual anchor such as the filling of a cylinder has an effect and can improve the quality of rating scale data that may improve a numerical comparison within and between respondents. We call for additional experimentation with similar anchors to study the strength of the effect and its applicability in a wider range of surveys that use rating scales.

References

Blasius, J./Thiessen, V. (2012). Assessing the quality of survey data. Sage.

Cunningham, W. H./Cunningham, I. C./Green, R. T. (1977). The ipsative process to reduce response set bias. Public Opinion Quarterly, 41(3):379–384.

Richins, M. L./Dawson, S. (1992). A consumer values orientation for materialism and its measurement: Scale development and validation. Journal of Consumer Research, 19(3):303– 316.

Rudnev, M. (2021). Caveats of non-ipsatization of basic values: A review of issues and a simulation study. Journal of Research in Personality, 93:104–118.

Schoonees, P. C./Van de Velden, M./Groenen, P. J. F. (2015). Constrained dual scaling for detecting response styles in categorical data. Psychometrika, 80(4):968–994.

Steenkamp, J.-B. E./Ter Hofstede, F./Wedel, M. (1999). A cross-national investigation into the individual and national cultural antecedents of consumer innovativeness. Journal of Marketing, 63(2):55–69.

Van der Heijden, P. G. M./De Falguerolles, A./De Leeuw, J. (1989). A combined approach to contingency table analysis using correspondence analysis and loglinear analysis. Journal of the Royal Statistical Society: Series C (Applied Statistics), 38(2):249–273.

Van Herk, H./Poortinga, Y. H./Verhallen, T. M. (2004). Response styles in rating scales: Evidence of method bias in data from six EU countries. Journal of Cross-Cultural Psychology, 35(3):346–360.

Van Rosmalen, J./Van Herk, H./Groenen, P. J. F. (2010). Identifying response styles: A latent-class bilinear multinomial logit model. Journal of Marketing Research, 47(1):157–172.

Part V: Reminiscences, anecdotes and greetings

Jörg Blasius – An Academic Life in Numbers
Miriam Trübner and Andreas Mühlichen

> "All data are dirty, but some are informative"
> (Blasius and Thiessen 2012)

A great deal of words would be necessary to describe our long friendships and working relationships with Jörg. There would be many endearing adjectives and if we were Shakespearian writers we would induce many tears in our readers – because we really, really like Jörg. A lot! Alas and woe, our writing is not entirely Shakespearian. So, instead, we would like to honor his academic life by sticking to what we have learnt best in our many years with Jörg as our mentor – reducing complexity with statistics and visualizations.

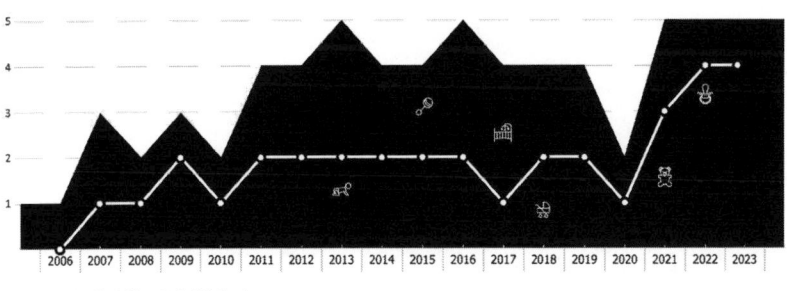

Additional Stats*	
research assistants (f)	16 (9)
department babies	6
secretaries	2
student assistants	16
PhD students (primary supervisor)	~20

Source: rather scetchy memory
* includes chair staff, staff of third-party funded research projects, and temporary staff due to parental leave

20 years of teaching in bubbles

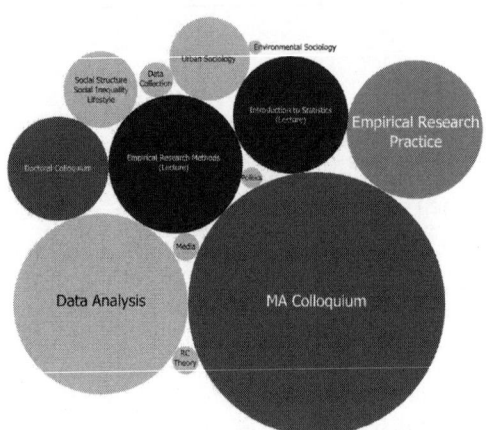

○ scale: 2 SWS
◐ seminar
◉ practical course
● colloquium
● lecture

Additional Stats

course types	~41
hours of teaching at Bonn University	~3224 hours[1]
hours spent in colloquiums	~1378 hours[1]
wine bottles killed in the doctoral colloquium	~490 (mean: 2,5; min: 2*; max: 7**)
research semesters	4

plus, various courses in summer schools, spring seminars, and on other occasions at different locations

[1] teaching hours with 45 minutes
* one red, one white
** recollections may vary depending on alcoholic intake

Souce: course catalogue (electronic, print version pre 2009) own calculations

Partners in ~~crime~~ publications

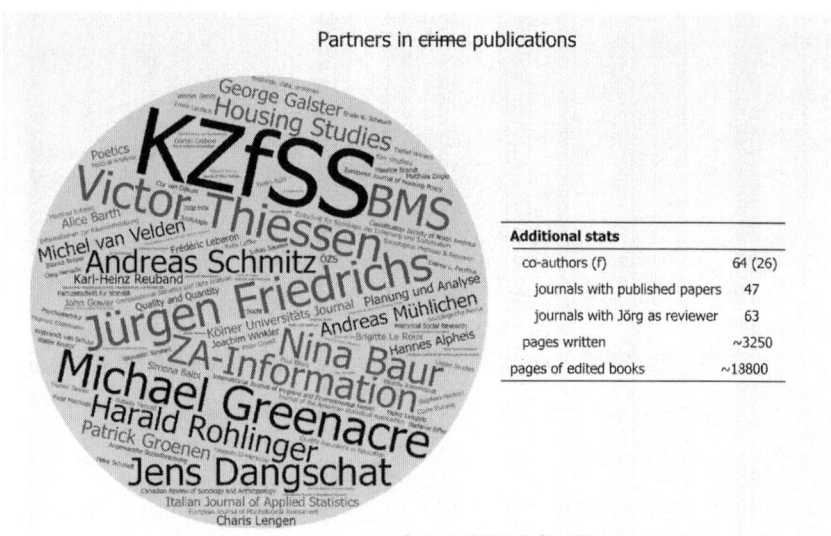

Additional stats

co-authors (f)	64 (26)
journals with published papers	47
journals with Jörg as reviewer	63
pages written	~3250
pages of edited books	~18800

Source: own calculations, data from website
https://www.politik-soziologie.uni-bonn.de/de/personal/prof-dr-joerg-blasius (18.01.2023)

Jörg's conference map

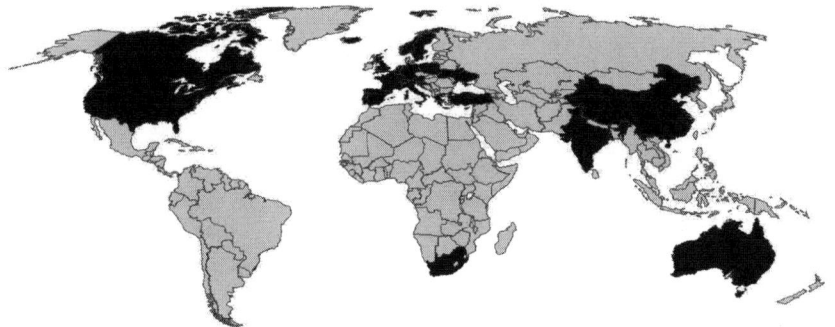

Source: https://www.politik-soziologie.uni-bonn.de/de/personal/prof-dr-joerg-blasius (18.01.2023)

Jörg's memberships

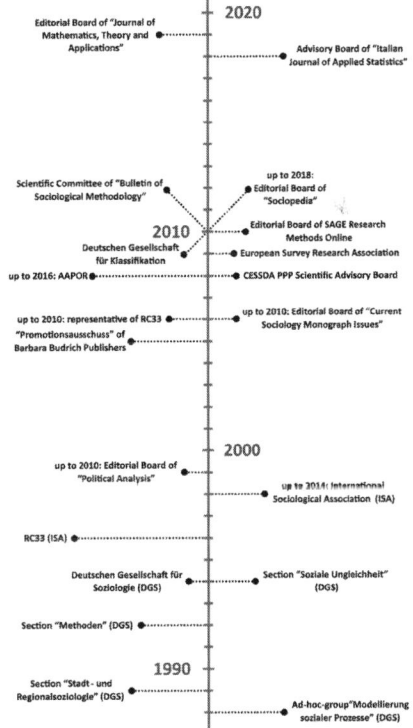

Source: https://www.politik-soziologie.uni-bonn.de/de/personal/prof-dr-joerg-blasius (18.01.2023)

Structure: follow culture!
Eine kurze Soziologie der akademischen Freundschaft
Clemens Albrecht

Zweigeteilt ist die Welt der modernen Wissenschaft: „Auf der *einen* Seite Wissenschaften mit dem Bestreben, durch ein System möglichst unbedingt allgemeingültiger Begriffe und Gesetze die extensiv und intensiv unendliche Mannigfaltigkeit zu ordnen. ... Der nie ruhende logische Zwang zur systematisierenden Unterordnung der so gewonnenen Allgemeinbegriffe unter andere, noch allgemeinere, in Verbindung mit dem Streben nach Strenge und Eindeutigkeit, drängt sie zur möglichen Reduktion der qualitativen Differenzierung der Wirklichkeit auf exakt meßbare Quantitäten ... Auf der *anderen* Seite Wissenschaften, welche sich diejenige Aufgabe stellen, die nach der logischen Natur jener gesetzeswissenschaftlichen Betrachtungsweise durch sie notwendig ungelöst bleiben muß. Erkenntnis der Wirklichkeit in ihrer ausnahmslos und überall vorhandenen qualitativ-charakteristischen Besonderung und Einmaligkeit".

Gesetzes- und Wirklichkeitswissenschaften, wie sie Max Weber hier auf den klassischen ersten Seiten seiner Wissenschaftslehre pointiert, nomothetische und ideographische Verfahren, wie sein neukantianischer Vorgänger Wilhelm Windelband in der berühmten Straßburger Rektoratsrede formulierte, sciences und humanities, wie Charles P. Snow die beiden Kulturen zu unvereinbaren Galaxien erklärte – solche binären Schemata beziehen sich auf Differenzen zwischen wissenschaftlichen Disziplinen. Manchmal werden sie auch innerdisziplinär gedeutet, etwa als ein Fortschrittsschema: von der Psychoanalyse zur modernen empirischen Psychologie, von der historischen Schule der Nationalökonomie über die Nomothetik der Menger'schen Grenznutzenlehre und die allgemeinen Gesetze der Neoklassik zur modernen, quantitativ-experimentellen Verhaltensökonomie, etc.. Das Narrativ: „von der geschichtsphilosophisch-kritischen Soziologie zur empirischen Kausalanalyse", von dem Moralsoziologen René König prolongiert und heute in der Gründung einer zweiten Fachgesellschaft institutionalisiert, entwickelt seine Plausibilität allerdings nur in der eigenen Stammeshälfte, und so bleibt die Soziologie als multiparadigmatische Disziplin Bürgerin in beiden Welten.

Zweigeteilt ist auch die Bonner Soziologie: Auf der einen Seite die Abteilung im Institut, auf der anderen das Forum Internationale Wissenschaft; auf der einen Seite die Methoden der quantitativen empirischen Sozialforschung in der Lennéstraße 27, auf der anderen die Kultursoziologie in der Lennéstraße 25. Getrennt sind diese beiden durch den Abstieg im einen, den Aufstieg im anderen Treppenhaus, vorbei an freundlichen und unliebsamen Stockwerktüren, alle gegen Eintritt gesichert durch Stahlgitter im Ordinarienkastell eines alten

Seminars für Politische Wissenschaft, das nie etwas anderes werden wollte. Organisationssoziologisch betrachtet könnte man mit allen Mitteln der Sozioprudenz eine Struktur nicht konfliktfreundlicher schaffen als sie hier bis in die Baulichkeit hinein institutionalisiert wurde – und in diesem Fall gilt leider die alte Ogburn'sche Erkenntis: culture follows structure.

Was tun, wenn man in eine solche Struktur berufen wird? Tenbruck lesen. In seinem klassischen Aufsatz zur Soziologie der Freundschaft beschreibt er, wie das aus der ständischen Gesellschaft entlassene Individuum eine neue Form der Festlegung von Person und Handlung in einer Interaktion fand, die sich durch Intensivierung frei gewählter persönlicher Nahbeziehungen bildete. Das ist der soziologische Gehalt des Freundschaftskults, der sich in allen europäischen Gesellschaften parallel zum Umbau von Stratifikation auf Funktion beobachten lässt: „In der persönlichen Beziehung entgeht der Mensch der Desorganisation, mit welcher ihn die Heterogenität seiner sozialen Welt bedroht."

Freundschaft allein überbrückt also die Klüfte, die sich zwischen Wissenschaftsidealen, Disziplinen, Erkenntnisinteressen und Methoden auftun. Auch sie findet ihre Formen zunächst im Organisationsfremden: in gemeinsamen Mahlzeiten beim Italiener, im Geselligen nach Disputationen, Einladungen zum Hauskonzert, im Habitus der stets fröhlichen Zugewandtheit. Ohne diese Informalität ließe sich die Konfliktstruktur des Formellen kaum konterkarieren. Und so bleibt im Mikrokosmos der Abteilung für Soziologie wie im Mesokosmos der Fächer oder im Makrokosmos der Gesellschaften die soziologische Grundeinsicht bestehen, dass alles Moderne, Rationale, Organisierte seinen heimlichen Kitt im Antiquierten, im Gefühlsmäßigen, im Informalen hat.

Lasst uns auch die organisatorische Struktur unseres Instituts diesem Prinzip anpassen: structure: follow the culture! – das ist zugleich der Auftrag für eine Fortentwicklung unserer wissenschaftlichen Erkenntnisse wie für die menschwürdige Einrichtung der sozialen Welt.

Persönliche Notiz
Nina Baur

Die Bedeutung von Jörg Blasius für meinen Werdegang kann gar nicht groß genug eingeschätzt werden. Persönlich kennengelernt habe ich Jörg 2004 auf der „Sixth International Conference on Social Science Methodology" des Research Committee on Logic and Methodology in Sociology (RC33) der International Sociological Association (ISA) in Amsterdam. Ich war eine Doktorandin von einer Provinz-Universität, in der es zwar eine sehr gute Methodenausbildung, aber keinerlei Anschluss an die Methoden-Community gab. Weil ich nicht nur zu einem methodologischen Thema promovierte, sondern auch schon damals wusste, dass ich in diesem Bereich bleiben wollte, war dieser Anschluss für mich sehr wichtig.

Meine ersten Erfahrungen bei dem Versuch, in der deutschen Methoden-Fachgemeinschaft Fuß zu fassen, waren eher schwierig, was vermutlich vor allem an den historischen Umständen lag: In Deutschland war der Streit zwischen qualitativen und quantitativen Methoden so stark eskaliert, so dass sich gerade die DGS-Sektion „Methoden der qualitativen Sozialforschung" aus dem „Arbeitskreis Qualitative Methoden" der Sektion „Methoden der empirischen Sozialforschung" ausgegründet hatte, was die Konfliktlinien erst zementierte – für jemand, für den Methodenintegration selbstverständlich und notwendig erscheint, war das eher ungünstig. Hinzu kam, dass ich zu jung, zu weiblich und zu unvernetzt war. Mir wurde von beiden Sektionen verdeutlicht, dass ich im Prinzip unerwünscht sei: Nach meinem Aufnahmevortrag bei der Sektion „Methoden der empirischen Sozialforschung" wurde ich nur mit einer Stimme Mehrheit aufgenommen, die Sektion „Methoden der qualitativen Sozialforschung" hatte komplett vergessen, dass ich bereits im Arbeitskreis und damit bei der Sektionsgründung langjähriges Mitglied war.

Da ich etwas störrisch bin und weiterhin unbedingt in den Methoden bleiben wollte, versuchte ich es auf internationaler Ebene – ich hatte von der RC33-Tagung gehört und sparte monatelang, um auf die Konferenz fahren zu können, und der Kontrast zu meinen Ersterfahrungen in Deutschland hätte nicht größer sein können. Ich traf eher zufällig auf Jörg, der damals designierter RC33-Vorsitzender für die nächste Amtsperiode war und mich sofort begrüßte. Er nahm mich unter die Fittiche, indem er sich genau anhörte, was ich machen wollte, mich Leuten vorstellte, mir Literaturhinweise und andere Tipps gab und mich ermutigte, aktiv am RC33-Geschehen teilzunehmen. Dieses herzliche, diskursoffene und inspirierende Klima war nicht nur Jörgs persönlicher Stil – sein Motto als RC33-President war: „Wer mitmachen und über Methoden diskutieren will, darf mitmachen!" –, sondern prägte das gesamte Research Committee – ich hatte eine wissenschaftliche Heimat gefunden. Bei der nächsten RC33-Tagung

2008 in Neapel – Jörg war nun RC33-President – reichte ich eine Session ein, aus der dann nach einem sehr erfolgreichen Call sechs Sessions wurden. Auf dem Weltkongress für Soziologie 2010 schlug mich Jörg als stellvertretende Vorsitzende für RC33 vor, zu der ich auch gewählt wurde, was nach RC33-Statuten bedeutet, dass ich automatisch als nächste Vorsitzende nominiert war. Alles Weitere – einschließlich der erfolgreichen Einwerbung des „Global Center of Spatial Methods for Urban Sustainability" (SMUS) wäre ohne Jörgs Förderung nicht möglich gewesen.

Parallel hatte mich Frank Engelhardt, der damals noch beim VS-Verlag arbeitete, gefragt, ob ich nicht zusammen mit ein oder zwei Mitherausgebern meiner Wahl ein „Handbuch Methoden der empirischen Sozialforschung" herausgeben wolle. Ich fasste mir ein Herz und fragte Jörg, der sofort zusagte. Die Kollegen mit einer Spezialisierung in qualitativer Sozialforschung, die ich kannte und gefragt hatte, hatten leider alle abgesagt, weshalb Jörg meinte „Lass uns das alleine machen – zu zweit arbeiten wir ohnehin viel besser!" Wir begannen die Arbeit am Handbuch 2010 – nach fünf Jahren war die erste Auflage endlich fertig und ist heute – obwohl es in deutscher Sprache erschienen ist – meines Wissens das meistverkaufte und -gelesene Buch von Springer weltweit und für alle Fächer. Wegen des Erfolgs ist geplant, das Handbuch in regelmäßigen Abständen zu verlegen – die dritte Auflage ist soeben erschienen.

Über die Jahre ist mir Jörg nicht nur Mentor, akademisches und persönliches Vorbild und guter Kollege gewesen, sondern auch ein sehr guter Freund geworden. Um konzentriert am Handbuch arbeiten zu können, hat sich eingespielt, dass wir bei Jörg zu Hause zwei- bis dreimal im Jahr für mehrere Tage in Klausursitzung gehen. In diesem Rahmen lernte ich auch Jörgs Frau, Beate kennen, mit der ich mich sofort hervorragend verstand, und auch wir sind mittlerweile sehr gute Freunde, weshalb wir bei den Besuchen immer extra Zeit einplanen

Mir bleibt daher nur zu sagen: Danke und alles Gute für die kommenden Jahre!

Zwei wichtige Erlebnisse – für Jörg und für mich
Jens S. Dangschat

Meine erste Begegnung mit Jörg, die ich noch gut erinnere, liegt schon über 40 Jahre zurück. Wir führten damals in den langen Semesterferien an der Forschungsstelle Vergleichende Stadtforschung der Universität Hamburg eine „selbstgestrickte" Summer School durch. In dem Zuge beschäftigten wir uns mit Modellierungen nach dem *system dynamics*-Modell von Jay Forrester, resp. dem Stadtregionen-Modell von Ira Lowry – auch *rational choice*, ganz nach dem Habitus der Soziologie an der Uni Hamburg, stand ganz oben. In diesem Kontext kam auch ein Student namens Jörg Blasius zum Zuge, der bei uns damals als studentische Hilfskraft beschäftigt war – intern als „Rechenknecht" bezeichnet.

Dieser Jörg Blasius wollte uns einen völlig neuen methodischen Zugang vorführen. Dabei ging es um die Verteilung von Zebras und Antilopen (oder waren es doch Elefanten?) im Krüger-Nationalpark! Bitte, was? Antilopen, Zebras, Afrika? Geht's noch? In der heftig folgenden Sinn-Debatte wurde Jörgs Stimme immer lauter, heller, sich immer wieder überschlagend – ja, das ist eine ganz neue multivariate Methode und gut geeignet für sozialwissenschaftliche Analysen – ihr werdet es sehen! Und der Bourdieu wendet diese Methode, die ein gewisser Greenacre in Südafrika weiterentwickelt hat, auch an! Dass Jörg längst mit Michael „auf Augenhöhe" ist und beide eng befreundet sind, das ist ja über die verschworene Korrespondenz-Analyse-Gemeinde hinaus bekannt.

Es gibt noch ein zweites sehr zentrales Erlebnis mit Jörg, denn er hat mir erst meine Karriere ermöglicht. Es war 1984, ich war an der TU Hamburg-Harburg in einem Projekt über die Hintergründe des Entstehens (eigentlich des Ausweitens) „neuer Haushalte" beschäftigt und wollte daraus über das Thema „Gentrification" bei Jürgen Friedrichs an der Uni Hamburg promovieren. Dann schmiss der damalige dem Schwerpunkt „Stadtsoziologie" zugeordnete Assistent schon nach drei statt sechs Jahren seinen Job hin, d.h. diese Stelle wurde überraschend früh frei. Nach einem Tag Bedenkzeit wurde der Urlaub gestrichen und das Promotionsthema zu „Segregation in Warschau" gewechselt. Ich hatte mich ja ein paar Jahre zuvor im Rahmen eines umfangreichen Stadtentwicklungsprojektes mit der Stadt Warschau befasst und aus der Zeit lagen zwei große Daten-Spulen mit bislang nicht ausgewerteten Volkszählungs-Ergebnissen aus zwei Zeitpunkten irgendwo in einem Schrank herum. Es gab aber kaum Informationen zu den Inhalten, wie genau die Daten abgelegt waren, etc.

Also ab ins Rechenzentrum und die Daten hochladen und mit SPSS 10 (?) für Großrechner auswerten. Leichter gesagt, als getan – zumal ohne „Rechenknecht". Wo war der Jörg? – Im Urlaub!!! Ich musste doch innerhalb eines Vierteljahres meine Diss abgeben!! Gott sei Dank kam Jörg nach einer gefühlten Ewigkeit gut erholt nach Hamburg zurück und machte sich ans Werk. Es hat geklappt, die Diss wurde zwei Tage vor Abgabetermin fertig und ich bekam die Assistenten-Stelle bei Jürgen. Jörg und ich haben in der Folge noch eine Reihe von Aufsätzen in amerikanischen Journals veröffentlicht. Das lief vor allem so gut, weil wir nachweisen konnten, dass es in einem sozialistisch geprägten Land deutliche Segregations-Phänomene gab – ein gefundenes Fressen auch noch in der auslaufenden Zeit des ‚cold war'.

Auch über die räumliche Trennung zwischen Köln und Hamburg, sowie Bonn und Wien haben wir den Kontakt miteinander nicht verloren und sind uns weiter freundschaftlich auch in mehreren Projekten verbunden geblieben.

An Jörg
Robert Helmrich

Lieber Jörg, wir kennen uns nunmehr seit rund 35 Jahren – also gefühlt eine Ewigkeit. Beide arbeiteten wir in den späten 1980er-Jahren in Köln in der Soziologie in der Greinstraße. Du warst im Vorderhaus bei Jürgen Friedrichs und später im Zentralarchiv auf der Bachemerstraße, und ich war im Hinterhaus der Greinstraße bei Erwin K. Scheuch beschäftigt. Die Soziologie hat uns dabei mehr räumlich im Sinne einer gelebten Soziologie als inhaltlich zusammengebracht. Der Soziologie-Stammtisch, der weder einen Stamm noch einen Tisch hatte und daher für uns auch nicht allzu spießig klang, traf sich monatlich auf Zurufe (Mails oder Whatsapp waren damals noch unbekannt), an wechselnden Orten und jeder, der es irgendwie erfuhr und sich angesprochen fühlte, kam dann dorthin.

Dies schaffte einen Zusammenhalt und eine Verbundenheit, die auf zwei sozialen Grundkonstanten fußte. Ein, wenn auch zugegebenermaßen sehr weites, aber dennoch gemeinsames thematisches Umfeld und das Streben nach sozialem Austausch darüber und auch darüber hinaus. Das thematische Umfeld war ein hinreichendes Kriterium um sich eingeladen zu fühlen. Es war aber nur der Einstieg, seltener der Anlass oder der Rahmen. Manchmal schloss es an einen Vortrag an, dann war es mitunter etwas steifer, aber dafür die Getränkefrage schneller geklärt. Manchmal ging man aber auch einfach gemeinsam ins Theater, oder in eine der Kölner-Brauhäuser oder zum Grillabend in den Volksgarten. Und so kamen anfangs nur eine Handvoll, später manchmal bis zu vierzig Personen. Es war eine Mischung aus kulturellem und sozialem Kapital, welches hier verbindend wirkte. Es war aber nie ein kölscher Klüngel, denn es erwuchs daraus kein ökonomisches Kapital. Seilschaften oder ähnliche Steighilfen wurden dort nicht gebildet, zumindest nicht in der breiten Fläche.

Was fast fünf Jahre so passierte und verband, verschwand aber genauso schnell wieder. Was sind die Bruchstellen für das Wirken solcher Grundkonstanten? Zeiten und Menschen.

Alles hat seine Zeit und ist diese vorbei, dann verändert sich auch das Umfeld. Aber wann endet die Zeit? Forschung und Lehre, Studiengänge und Seminare unterliegen einem Zeitplan, der auf Veränderungen ausgerichtet ist. Zieldefinition und Zielerreichung sind darin mehr oder weniger klar abgebildet und die zeitlichen Beschränkungen sind allen Beteiligten stets bewusst. Zurück bleibt nur ein kleiner Stamm von Personen, die institutionell gebunden sind und die Universität zumindest für eine vergleichsweise längere Zeit definieren. Selten sind dies aber die Personen, die die Aktivitäten hervorbringen.

Es sind die Menschen, auf die es ankommt. Wie in allen sozialen Einheiten, gibt es auch hier diejenigen, die die Initiative ergreifen und diejenigen, die als Statisten mitwirken. Während die Aktivisten aber ohne Statisten weitermachen können, verfallen die Statisten ohne die Aktivisten zumindest in diesem Kontext in eine anhaltende Starre.

Warum sind solche informellen Austausche bedeutsam? Sie bilden Lebensstile aus und schaffen soziale Mobilität. Denn diese Treffen und informellen Strukturen bilden Austauschprozesse des ökonomischen, kulturellen und sozialen Kapitals für die Bildungs- und Erwerbschancen von jungen Menschen. Die jungen Menschen, die sich im Studium befinden oder ihre ersten Versuche im Erwerbsleben machen benötigen den sozialen Halt und eine soziale Orientierung, die über das Elternhaus hinweg wirkt.

Du hast, wie ich auch, Dein Wirken nach Bonn verlagert. Und da hast Du Deine eigene Art des sozialen Zusammenhalts entwickelt. Doktorandenseminar mit Wein, Hauspartys und Houseconcerts als offene Räume mit einem sinnlichen Ambiente für den sozialen, wissenschaftlichen, kulturellen und vor allem musikalischen Austausch. Mal sehen, was da alles noch so kommt. Denn auf die aktivierenden Menschen in den sozialen Räumen kommt es an.

An Jörg
Rahim Hajji

Lieber Jörg,

ich kann mich noch sehr gut an dein Vorsingen erinnern. Ich bin damals zum ersten Mal mit der Korrespondenzanalyse in Berührung gekommen. Ein Verfahren, das mich bis heute beschäftigt.

Nach deinem Ruf an die Universität Bonn habe ich als Student, Tutor und wissenschaftliche Hilfskraft bei Dir viel gelernt. Unter anderem habe ich unterschiedliche statistische Verfahren kennengelernt. Darüber hinaus konnte ich bei dir erste Erfahrungen als Lehrender sammeln, die meine Lehre bis heute prägen.

Als Doktorand habe ich mit einer qualitativen Studie zu Deutsch-Marokkaner/innen bei dir promoviert. Im Rahmen der Dissertation genoss ich die Handlungsspielräume bei Dir sehr. Auf diese Weise gabst Du mir die Möglichkeit, mich persönlich und wissenschaftlich weiter zu entwickeln.

Deine Offenheit, deinen Pragmatismus und deine Neugierde für empirische Methoden und Themen jeglicher Art bewundere ich bis heute.

In tiefer Dankbarkeit,

Rahim

Für Jörg
Yasemin El-Menouar

Lieber Jörg, wir sind uns das erste Mal begegnet, als ich – damals als Studentin der Soziologie an der Universität zu Köln – in Deiner Statistik-Vorlesung saß. Anschließend fragtest Du mich, ob ich Lust hätte deine Statistik-Kurse als Tutorin zu begleiten und stelltest mich als studentische Hilfskraft bei Dir ein. Das ist jetzt etwa 25 Jahre her. Und was soll ich sagen? Einen besseren Start ins Berufsleben hätte ich mir nicht vorstellen können. Von Dir habe ich nicht nur meine Begeisterung für statistische Methoden – insbesondere die Korrespondenzanalyse – gelernt, Du hast mich auch als Mentor gefördert, und als guter Freund und Vertrauensperson immer unterstützt.

Ich kann mich noch sehr gut an unsere gemeinsame Zeit in Essex erinnern, als wir beide an der Summer School for Social Data Analysis teilnahmen – ich als Studentin, Du als Dozent. Auch Michael Greenacre war damals dort. Diese intensive, lehrreiche Zeit, in der auch das Feiern nie zu kurz kam, hat mich sehr geprägt. Es war nicht selbstverständlich, dass ich an diesem dreiwöchigen Seminar teilnehmen durfte. Vielmehr war es Deinem Engagement zu verdanken, dass ich die Förderung dafür erhielt.

Seitdem blicke ich auf zahlreiche gemeinsame Frühjahrsseminare bei GESIS – damals ZA – zurück, teils als Deine Teaching Assistant, später auch als Mitarbeiterin in Bonn und im DFG-geförderten Projekt „Leben in benachteiligten Wohngebieten", das du gemeinsam mit Jürgen Friedrichs geleitet hast, gemeinsame Konferenzen – insbesondere die legendären CARME-Konferenzen, aber auch auf die vielen Hauspartys, unsere beiden Hochzeiten, die Geburten unserer Kinder und Deiner Enkel zurück.

„Nächste Woche habe ich Urlaub – und endlich Zeit zum Arbeiten" – dieser Satz ist mir besonders in Erinnerung geblieben. Oder auch Deine Begeisterung für Flug- und Fahrpläne. Nicht zu vergessen: deine indischen Kochkünste – hmm, der Curry-Duft liegt mir noch jetzt in der Nase. Oder auch deine Fähigkeit, jeden Hauch von Antriebslosigkeit in der Runde mit einem energischen „Let's go" in Motivation zu verwandeln.

Vielen Dank für die lehrreiche Zeit – sowohl professionell als auch menschlich – sowohl als Mentor und Vorbild, als auch als Freund.

An Jörg
Martin Fritz

Lieber Jörg,

im Jahr 2001 habe ich als studentische Hilfskraft im Zentralarchiv für Empirische Sozialforschung in Köln (ZA) angefangen zu arbeiten. Da warst Du schon wieder weg und über die Station Darmstadt auf dem Weg nach Bonn, um Deine Professur anzutreten. Trotzdem habe ich Dich durch die Arbeit am ZA kennengelernt. Denn viele Kolleg:innen wie Franz, Maria und Ingvill kannten Dich und berichteten von Dir und Deiner Arbeit. Von Korrespondenzanalyse verstand ich damals noch nichts – ich hatte zu der Zeit Philosophie studiert und Soziologie war noch nur eine Nebenbeschäftigung, von Statistik ganz zu schweigen. Als ich später in jenem Jahr mit Yasemin zusammenkam, lernte ich Dich dann auch persönlich kennen. Ich weiß nicht mehr genau bei welchem Anlass es war, wahrscheinlich bei einer der zahlreichen und immer sehr schönen Hauspartys in Hilden. Ich erinnere mich noch, dass einer Deiner internationalen Gäste mich gefragt hatte, was ich arbeite. Mein Englisch war damals noch äußerst rudimentär, so dass ich nur „I'm working in an office" antworten konnte. Vielleicht lag's aber auch teilweise an dem immer zu diesen Gelegenheiten großzügig ausgeschenkten Rotwein. Jedenfalls waren die Partys, die ihr später in Bonn weiter veranstaltet habt, jedes Mal ein schönes soziales Ereignis, bei dem wir viel Spaß hatten.

Es hat dann einige Zeit gedauert bis wir uns auch in einem Arbeitskontext häufiger gesehen haben. Im November 2010 habe ich offiziell meine Promotion an der Uni Bonn mit Dir als Erstbetreuer angemeldet und etwas später begonnen bei Dir am Institut auch Lehre anzubieten, die Übungen zur Einführung in die Statistik sowie Kultur und Sozialstruktur. Statistik selber zu lehren hat einige Vorbereitung erfordert und mein Verständnis des ganzen deutlich verbessert – darüber eine Anbindung an den Lehrstuhl zu haben, hat mir auch ein gutes Gefühl gegeben und geholfen, meine Promotion voran- und zu Ende zu bringen. Am Promotionskolloquium konnte ich leider nicht so häufig teilnehmen, da ich in Köln wohnte und Yasemin und ich mittlerweile viel Zeit mit unseren beiden Kindern verbrachten (Ayla war schon 2007 auf der CARME Konferenz in Rotterdam dabei). Die wenigen Termine beim Promotionskolloquium waren umso wichtiger und hilfreicher für mich, denn in der entspannten Atmosphäre unter dem Dach des Altbaus, in dem sich die Räumlichkeiten Deines Lehrstuhls befanden, ließen sich alle Inhalte und Fragen wunderbar besprechen. Dafür lohnten sich die Zugfahrten von Köln-Mülheim nach Bonn immer und in etwas weniger als schlappen sechs Jahren war die Dissertation auch fertig und abgegeben.

Der Beitrag für diese Deine Festschrift, den Yasemin und ich zusammen verfasst haben, trägt den Titel „Religiosity and climate change beliefs in Europe" und kombiniert unsere beiden Themen, mit denen wir uns beruflich beschäftigen. Bei der Recherche dazu sind wir auf ein ganz eigenes und spannendes Forschungsfeld gestoßen – die Zusammenhänge zwischen Religion und Vorstellungen vom Klimawandel und der Natur. Das hat auf jeden Fall noch weitere und detailliertere Analysen mit CARME verdient und wer weiß: vielleicht werden wir das in Zukunft gemeinsam angehen. Es wäre schön, Dir darüber auf einer der hoffentlich noch vielen kommenden Hauspartys zu berichten.

Travels with Jörg
Patrick J. F. Groenen

Over much of my professional life, Jörg has been a constant friend. For me that started with probably one of the first predecessors of the CARME meetings: the 1991 meeting at the University of Cologne called Correspondence Analysis in the Social Sciences. There, I gave one of my first international conference presentations. The topic was on the ANACOR program for correspondence analysis that the Leiden group had ported to SPSS. After that, many more travels with or to Jörg happened. I have spent many happy days at his home. Coming from the Netherlands, first Hilden and later Bonn became a not-to-miss stop on my holidays for me and my family. At one of these occasions, on route to Turkey in the winter through Düsseldorf Airport, I remember getting stuck in a blizzard: the abundance of snow closed Dusseldorf airport and with the last available taxi, we arrived at Jörg's place warm and cosy, where we waited for the snow to retreat.

Figure 1. Some examples of my travels with Jörg

a.  b.  c.

Panel a is Patrick, Jörg, and Beate in Hilden, Panel b at the Çanakkale ferry in 2012, and Panel c is at the Carme 2003 meeting in Barcelona with Michael Greenacre, Yasemin El-Menouar, Jörg Blasius, Tor Korneliussen, and John Gower.

Jörg likes to travel and travel to conferences and workshops. There is an almost endless list of conferences that he has been to and a sizeable set of conferences that we both visited. Figure 1 shows three photos of traveling to or with Jörg.

His house has been an open and welcoming place for many friends and guests. Well known have been the in-house concerts with sometimes 60–80 visitors and music styles ranging from subtle as by Karol Green to rough by a local punk band. Many colleagues and friends have come to these special events.

The lockdowns during the 2020–2022 pandemic was something that Jörg did not like at all. It turned out that at the end of February 2020, I was the last visitor at Jörg's home before the pandemic broke out. It was more than one year later, early July 2021, that I visited Jörg again and it turned out that I was the first to return after the lockdowns. For me, it was wonderful being back after that uncertain year.

The travels are not over and will not be over. During Easter, 2022, we decided to meet up near Barcelona. The weather turned out to be exceptionally warm for the time of the year and had a beautiful time with Jörg, Beate, Michael, and Hilde. All of us were careful at the time to avoid catching Covid. However, one day after my return to home, I tested positive. In the days that followed, four out of the five on this trip also tested positive. It will remain a mystery who infected who, and I do not care: the travels with Jörg always have been a great pleasure and that will remain so.

My 39 years with Jörg
Michael Greenacre

In 1984 Prof. Walter Kristof invited me to give a two-day course on correspondence analysis at his sociology department in Hamburg, when I was next in Europe. I had already met Walter a few times in South Africa, when he visited my colleague Michael Browne in our statistics department at the University of South Africa in Pretoria. I accepted the invitation and, when in Germany, stayed with Walter and his family in Hannover for a couple of days. On the first day of the course, which was a Thursday, we drove to Hamburg and I remember that Walter let me drive very fast for a short time in his fast car on the unlimited speed highway, and I did about 160 km/hour, the fastest in my whole life. On the way, Walter warned me that there was going to be a student demonstration against my giving a course, since I came from South Africa and there was a cultural and scientific boycott against South Africa at that time.

Sure enough, when we arrived at the lecture hall, which was supposed to contain only six post-graduate sociology students, there were about 60 more students, who immediately started to protest about my presence and insisted that I could not give the course. A long debate started between Walter and these students, in German, which I couldn't understand, so I started up a conversation in English with one of the protesters sitting next me. We had a lot in common because I was also against the South African government at the time and its policies. After about 45 minutes Walter was getting angrier and angrier at the stalemate in the class, and he called us, the six students and myself, to go to his office. The verbal bantering now turned more physical, as the protestors followed us to the office door and insisted to enter as well. There was a lot of pushing and shoving and eventually we managed to get into the office, and while Walter was pushing the door against the protestors, one of them managed to get inside!

With the door locked, we all tried to settle down. I was quite rattled by this experience and I asked for some minutes to recover. I remember looking out of the office window and thinking about another incident that was very much in the news just then. This was about the South African middle-distance runner Zola Budd, competing in the Olympic Games in Los Angeles in August, being involved in a very controversial incident, where she and the other favourite for the 3000m finals bumped into each other more than once, affecting both their chances of winning. Because of the time of the year, I estimate I was giving this course in Hamburg in September 1984.

Just to finish off this part of the story, I pulled myself together, and then – using only a blackboard – I started to give my course. After about an hour of introducing categorical data and some matrix algebra, the protestor, apparently a first-year political science student, who was sitting quietly during my lecture, now very much in the minority, asked if we could unlock the door and let him out! My course continued that day, and when we emerged from the office for lunch there was no-one outside and it continued without any more incidents until Friday afternoon. On Friday we came back to the original room – as Walter said, the protestors would not come back on a Friday! About a month later, back in South Africa I received a letter of apology from the Dean about this incident, and I also received from Walter two pamphlets that were circulating around the university: one denouncing my presence at the university and another one supporting my right to give the class!

Anyway, why do I explain all this? It's because Jörg was one of those six students, aged 27. It is impossible to forget him with his very long hair. And he was very interested in the material on correspondence analysis which I presented. He eventually did his doctorate on this subject with Jürgen Friedrichs and Walter as his supervisors, and he used correspondence analysis in his thesis.

**Jörg Blasius,
as a young post-graduate sociology student**

Original photo by Ines Jarchow,
restored by Françoise Greenacre

Now wind forward from 1984 to 1990/91. I was on sabbatical at the *École des Mines* in the *technopole* of Sophia Antipolis, south of France, when Walter contacted me again. He told me that his former student Jörg was now working in Cologne at the *Zentralarchiv für Empirische Sozialforschung*, and he proposed that Jörg and I organize an international meeting there on correspondence analysis. Thanks to the generous support of the Archive, directed then by Ekkehard Mochmann, who had a strong confidence in this project, we did it, in April 1991. And we did it again in 1995 and 1999, sponsored again by the Archive, thanks once more to Ekkehard's support. In 2003 we organized it in Barcelona and called the conference *CARME* (Correspondence Analysis and Related Methods), which is a Catalan girl's name. And then Rotterdam in 2007, Rennes, France in 2011 (the 50th anniversary of Benzécri-style correspondence analysis), Napoli, Italy in 2015, then outside Europe for the first time in Stellenbosch, South Africa in 2019, and now for the ninth time in Bonn, coinciding with Jörg's retirement. Jörg and I turned CARME into a well-known and friendly meeting on a wide range of multivariate methods, more or less centred around the analysis of categorical data. We edited four books as well as two special journal issues – see the bibliography below. The books, which are not conference proceedings but each on a particular theme and with invited contributors, have received excellent reviews.

Apart from these academic activities, Jörg and his wife Beate have become best friends in more ways than one – I was Jörg's witness at their marriage, and they were witnesses as well at my marriage, to Zerrin Aşan in Turkey. I know Jörg for 39 years and it has been a story of great friendship, hard work together on CARME and the publications, and many, many fun times as well. We also did a trip to the USA together to give a course at the Gallup Research Center in Lincoln, Nebraska, and then had an unforgettable experience in Chicago going to *Buddy Guy's Legends* jazz club to listen to the music of Bernard Allison. I wish him all the best for his retirement in the years ahead with Beate, enjoying their grandchildren, nice trips around the world and more good times together with their many friends.

Bibliography by Blasius and Greenacre

1994: Greenacre, M. & Blasius, J. (eds), Correspondence Analysis in the Social Sciences: Recent Developments and Applications, Academic Press, London.

1998: Blasius, J. & Greenacre, M. (eds), Visualization of Categorical Data, Academic Press, New York.

2006: Greenacre, M. and Blasius, J. (eds), Multiple Correspondence Analysis and Related Methods, Chapman & Hall/CRC Press, Boca Raton, USA.

2009: Blasius, J., Greenacre, M., Groenen,, P.J.F., & van de Velden, M. Special Issue on Correspondence Analysis and Related Methods, Computational Statistics and Data Analysis 53, 3103–3106.

2014: Blasius, J. & Greenacre, M. (eds), Visualization and Verbalization of Data. Chapman & Hall/CRC Press, Boca Raton, USA.

2017: Balbi, S., Blasius, J. & Greenacre, M. Special Issue on Correspondence Analysis and Related Methods – CARME, Italian Journal of Applied Statistics 29, 117–123.

Eine Annäherung an Dich Jörg, an uns Reisende

von Wendelin Strubelt

Wir sind nicht am Ende einer langen Reise.
Noch nicht.
Wir sind nicht die einzigen Reisenden.
Viele gingen uns voraus.
Viel gerühmte wie Odysseus.
Besungen vor langer Zeit.
„Nenne mir, Muse, die Taten des vielgewanderten Mannes,
Welcher so weit geirrt, nach der heiligen Troja Zerstörung......"
auf seinen Irrfahrten in der damals bekannten Welt.
Aber -
Wo bin ich, sind wir hier?
Was hat Troja mit Dir, mit uns zu tun?
Solche Zerstörungen und ihre Kämpfe liegen weit hinter uns
oder doch nicht?
Wie war es da, damals, wo Du herkommst?
Wer und was brachte Dich hierher?
Welche Muse ist hier gemeint,
welche paßt auf Dich?
Es kann nur Kalliope sein -
die Muse des Nachdenkens, des Schreibens -
vielleicht auch die der Zahlen
hinter den Worten,
warum denn nicht auch?

Aber halt:
Später, sehr viel später
näher an uns
nur vor hundert Jahren
die Irrgänge des Leopold Bloom

in einer Stadt an nur einem Tag
auf den Spuren des jetzt so genannten Ulysses
kein Krieger, eher einer wie Du und ich.
Aber mit einem anderen Anfang:
*"Gravitätisch kam der dicke Buck Mulligan vom Austritt am oberen
Ende der Treppe......"*
Aber so erscheinst Du nicht,
wenn wir uns treffen.
Du bist weder Odysseus
noch in römischer Gestalt Ulysses
oder in der irischen Variante
nun Leopold Bloom genannt,
der moderne Wanderer jetzt nur durch einen Tag,
der am Anfang nicht einmal genannt wird.

Doch auch Du kommst vom Meer,
nicht aus Ithaka, nicht aus Dublin,
aus einer Stadt an einem Fluß
nur in Meeresnähe.
Wir trafen uns nicht dort,
begegneten uns im Nachdenken über Städte –
ihren Abbildern in Zahlen.
Du fragtest nach deren Gewebe, dem Muster des Zusammenhalts,
auch Struktur genannt
und bliebst dabei.
Aber viele Fragen immer noch offen
trotz all der Hypothesen, die klären sollten
die variierenden Spielarten
der Beziehungen zueinander
und auch untereinander
auch Variablen genannt -
im verflochtenen Diskurs der
Befunde aus aller Herren Länder.
Alles weniger geleitet von Musen
den durchaus weiblichen Wesen,

eher geprägt von Hohepriestern auf ihren Stühlen
erschriebenen oder selbst besetzten.

Auf jeden Fall jetzt nicht mehr
getrieben vom Zorn der Götter oder eines Gottes.
Auch nicht mehr herumirrend
in und um Gegenden unwirtlicher Art
geschweige denn in Wirtshäusern.
Aber eben auch wieder in der Suche
nach:
Ja nach was, nach
sozialen Tatbeständen –
ein Wort nicht den Büros entflohen –
und nach bindenden Fäden
Gespinsten – nicht Gespenstern –,
die allerorten gesponnen werden,
um zu finden und
zu erklären das Geflecht,
die vorhandenen Verflechtungen.
Allemal als kompliziert bekannt –
gleichwohl, um Verstehen zu erleichtern
für uns.
Leider nicht für jeden, geschweige denn für alle.
Wer die auch immer sind.

Also wieder keine geraden Wege und
klare Antworten zu erwarten
auf den Wegen des Lebens
vom Norden in die alte Mitte
am Rhein.
Und hier nun ein neuer Einschnitt,
keine Kehre, eher ein Aufbruch.
Ein neuer Anfang für Dich

Lieber Jörg,
kein Herumirren wie die anderen
berufenen Gestalten von einst –
eher ein Abrunden von Gedanken und Leben
nach vollbrachter Pflicht.
Um in den Bildern der Alten zu bleiben:
Das Setzen neuer Segel
mit den Winden von damals wohl kaum.
Aber wieder ein Ankreuzen
gegen – gegen was denn nun?
Wohl kaum von Ithaka in die irischen Pubs
und wieder zurück?
Eher vom Katheder in die Weiten,
jenen fernen und jenen nahen
gleich um die Ecke..........

Der Griff nach den Musen
könnte Dir, uns helfen
„...auf all unseren Wegen"
nicht die Orientierung zu verlieren.

Aber nun ein Ende, nicht noch mehr Rückgriffe
auf weitere Wortgeflechte der Menschen
aus dem „stream of consciousness"!
Nur der Wunsch
"ad multos annos"
und halte die Ohren steif, alter Freund,
stürmische Winde gibt es allerorten
und immer wieder.
Aber: Du bist nicht allein,
ob in Ithaka, Dublin oder Bonn
oder in noch anderen Ecken der Welt.
Wir werden um Dich sein.

Author information

Clemens Albrecht

Clemens Albrecht is professor of cultural sociology at the University of Bonn. His main research interests are cultural sociology, history of sociology, and socioprudence.

Rebekka Atakan

Rebekka Atakan works as a PhD student in an DFG-funded project on the analysis of neighborhood changes by means of a dwelling panel. Her dissertation focuses on the perception of urban space, the construction of neighborhood images and the interplay of social and spatial inequality.

Alice Barth

Alice Barth works as a post-doc researcher in the department of sociology at the University of Bonn and currently leads a DFG-funded project on the analysis of neighborhood changes by means of a dwelling panel. Her main research interests are inequality, change in urban space, multivariate analysis methods, comparability of measurement instruments and relational sociological theories.

Nina Baur

Nina Baur is Professor for Methods of Social Research at Technische Universität Berlin and principal investigator in the subproject "Knowledge and Goods II" (A03) of the Collaborative Research Center "Refiguration of Spaces" (CRC 1265) funded by the German Research Foundation (DFG). Her research fields include social science methodology, process sociology, spatial sociology, and economic sociology (including sociology of consumption and sociology of work).

Jörg Breitung

Jörg Breitung is Professor for Econometrics and Statistics at the University of Cologne. His research fields are the analysis of time series and panel data in empirical macroeconomics and finance.

Jens S. Dangschat

Jens S. Dangschat is professor emeritus at the Technische Universität Wien. As an urban sociologist, his main research interests are socio-spatial inequalities (gentrification; segregation), sustainable spatial development and the effects of future technologies on (urban) society.

Rainer Diaz-Bone

Rainer Diaz-Bone is professor of sociology at the University of Lucerne. His main research areas are philosophy of science, social research methods and statistics, sociology of social research as well as neopragmatism and neostructuralism (applied to the analysis of culture and economy).

Aykiz Dogan

Aykiz Dogan is a political sociologist, specialized in the historical sociology of the State in the contemporary period. She currently works at Université Panthéon-Sorbonne, IEDES and UMR D&S. She has recently defended a thesis on the formation of the Turkish state in the inter-war period, and published various articles on the group of central bankers and its transformations since the end of the 1990s.

Yasemin El-Menouar

Yasemin El-Menouar is sociologist and scholar of Islamic studies. She is Senior Expert at the Bertelsmann Stiftung in Germany and scientific leader of the "Religionsmonitor".

Martin Fritz

Martin Fritz is a Post-Doc researcher in the BMBF junior research group "Mentalities in Flux. Imaginaries and social structure in modern circular bio-based societies" (flumen) at the Institute of Sociology of the Friedrich Schiller University Jena. His research interests include international comparative survey research, social and environmental attitudes and value orientations, human-nature relations, degrowth and sustainable welfare.

Michael Greenacre

Michael Greenacre is Senior Talent Professor at the Department of Economics and Business, Universitat Pompeu Fabra, Barcelona, also teaching in the Barcelona School of Management. His research interests are in multivariate statistics, especially correspondence analysis and, more recently, compositional data analysis in the fields of bio- and geochemistry, as well as genomics and microbiomics.

Patrick J.F. Groenen

Patrick J.F. Groenen is a professor of statistics at the Erasmus School of Economics (ESE). He currently is also dean of that school. His work focuses on data science techniques and their numerical algorithms.

Rahim Hajji

Rahim Hajji works as a sociologist at the University of Applied Sciences Magdeburg-Stendal. He teaches in the field of empirical social research and conducts research on migration, integration, education and health issues.

Robert Helmrich

Robert Helmrich is head of division and vocational education and training (VET) researcher in the area of "Qualification, Vocational Integration and Employment" at the Federal Institute for Vocational Education and Training (BIBB). His research focuses on skill developments and skill requirements in the labor market, structures and developments of occupations, job characteristics, skills, industries and firms, employment structures and opportunities in the labor market, new technologies, new production processes and green economy. He is also honorary professor of sociology at the University of Bonn.

Elmar Kulke

Elmar Kulke is Professor for Economic Geography at Humboldt-Universität zu Berlin. He is principal investigator in the subproject "Knowledge and Goods II" (A03) of the Collaborative Research Center "Refiguration of Spaces" (CRC 1265) funded by the German Research Foundation (DFG) as well as co-speaker of the German Working Group on Geographical Retailing Research. His research fields include locational changes in retailing/enterprise-oriented services and the development of global commodity chains, especially for fresh food articles.

Frédéric Lebaron

Frédéric Lebaron is a sociologist working at Université Paris-Saclay and ENS Paris-Saclay, IDHES, specialized in economic sociology and methodology. He has published various books and articles on the fields of economists, central bankers and economic policy makers using Geometric Data Analysis as an approach allowing to investigate social spaces. He has also published handbooks and articles in quantitative methodology for social sciences and participated to the study of global social inequality based on cultural, economic and political dimensions.

Brigitte Le Roux

Brigitte Le Roux is a mathematician, specialized in Geometric Data Analysis. She works at Université Paris-Cité, MAP5, and Sciences Po, CEVIPOF. She has recently published a reference book in the field of combinatorial inference applied to Geometric Data Analysis. She is also the author of reference books in the fields of GDA and statistics, in English and French.

Felix Leßke

Felix Leßke is currently employed as an evaluator at the German Development Cooperation Evaluation Institute (DEval). Previously, he worked as a research assistant at the Universities of Bonn and Cologne. His research focuses on empirical social research, migration and integration research, relational sociological theory, and urban sociology.

Andreas Mühlichen

Andreas Mühlichen works as an associate researcher at the University and City Library of the University of Cologne for the Cologne Competence Center for Research Data Management (C^3RDM). He offers consultations and trainings on all things RDM and has a background in quantitative empirical social research.

Simone Pollak

Simone Pollak works as a research assistant at the Magdeburg-Stendal University of Applied Sciences in the Department of Social Work, Health and Media.

Jessica Schäfer

Jessica Schäfer, M. A., is currently working as a researcher for the Magdeburg-Stendal University of Applied Sciences. As part of the digitization project h2d2, she works closely with the MaSta-Lab and the two XperiMaker labs in Magdeburg and Stendal as an accompanying researcher. Based on the concept of the DBIR approach, her goal is to advance the digitization of teaching as well as to establish innovative teaching-learning concepts at the h2.

Yvonne Scheit

Yvonne Scheit works at the Department of Sociology at the University of Bonn. Having a background of both Cultural Anthropology and Sociology, she is interested in multivariate analysis, qualitative methods, grounded theory and mixed methods. Her current research focuses on students' view of the digitalization process at the University of Bonn and the effects of the covid-19 pandemic on students' lives and digital teaching.

Manuela Schmidt

Manuela Schmidt works as a research assistant for the DFG-funded project "The Cologne Dwelling Panel" and is a PhD candidate of sociology at the University of Bonn. Her research interests include survey methodology, computational social science, and multivariate analysis methods.

Pieter C. Schoonees

Pieter C. Schoonees is a statistician and lecturer affiliated with the Econometric Institute of the Erasmus School of Economics in Rotterdam. His expertise lies in computational statistics and machine learning with an emphasis on cluster analysis and dimension reduction.

Ulrike Scorna

Ulrike Scorna (M.A.) is a social scientist and works as a researcher at the University of Applied Sciences Magdeburg-Stendal in the project ZAKKI (Central Contact Point for Innovative Teaching and Learning of Interdisciplinary Competencies of AI) in the field of accompanying research on conditions for success in teaching and learning of AI competencies. Her research focuses on digital assistive technology and medical sociology.

Wendelin Strubelt

Wendelin Strubelt (born 1943), Dr, Social Scientist. 1973 – 1980 University of Bremen (Assistant Professor, Professor, Vice Rector), 1980 – 1998 Director and Professor of the Federal Research Institute for Regional Geography and Regional Planning, Bonn. After fusion with the Federal Building Office until retirement 2008 Vice President and Professor of the Federal Office for Building and Regional Planning. Since then, Lecturer at the University of Bonn. Member of the German Academy for Urban and Regional Spatial Planning (DASL) and of the Academy for Territorial Development in the Leibniz Association (ARL).

Miriam Trübner

Miriam Trübner works as a post-doc researcher at the department of Sociology and Quantitative Social Research Methods at the Johannes Gutenberg University of Mainz. Her current research centers around household behavior (work, consumption, health, nutrition) in the context of social inequality and differentiation. She also has a particular interest in quantitative research methods and data quality.

Michel van de Velden

Michel van de Velden is an associate Professor of Statistics at the Econometric Institute of the Erasmus University Rotterdam. His main research interests are exploratory data analysis; in particular, dimension reduction and cluster analysis methods with a strong focus on data visualization. In addition, he is involved in several supervised machine learning projects involving tree-based machine learning methods.

Hester van Herk

Hester van Herk is professor of Cross-Cultural Marketing Research at the Vrije Universiteit Amsterdam and Adjunct Senior Research Fellow at the University of Western Australia in Perth. Her main research interests are in the antecedents and consequences of human values on consumer behavior in developed and emerging markets and in research methodology providing insight into differences and similarities between survey responses from consumers in different nations and cultural groups.

Karl M. van Meter

Karl M. van Meter has dual Franco-American nationality and university degrees in social sciences, mathematics and French literature from three different countries. He worked at the Sorbonne University, at the School of Advanced Studies in Social Sciences, at the CNRS and was "Invited Scholar" in the United States and Germany. He is currently at the Maurice-Halbwachs center of the Normal School Superior of Paris. He has widely published books, book chapters and scientific articles in French and in English that you can find under his name on Google-Scholar, ResearchGate and many other websites.

Gunnar Voss

Gunnar Voss is a sociologist and works as a researcher at Magdeburg-Stendal University of Applied Sciences. His main research interests include education studies, digitization, health inequalities as well as refugee and migration studies. In his work, he applies quantitative and qualitative social research methods including the approach of design based implementation research.

Index

B
Bourdieu 8–9, 47–48, 50, 52, 71–82, 132, 138–140, 164, 169–170, 208, 234

C
categorical principal component analysis 8–12, 21–28
category weights 27–28
climate change 8–12, 83–105, 84–105, 140–148, 239–256
Combinatorial Inference in GDA 46
correspondence analysis 7–10, 13–28, 29–45, 46–48, 69–82, 84, 91, 101, 155, 159–160, 208, 214, 221–223, 240–244

D
data worlds 10, 195–212
distance measures 30, 35, 40–41, 45

E
eating 9, 164–194
ECB 46–62
economics of convention 196
educational inequality 106–130

F
food markets 165, 188, 190

G
GDR 133, 149, 153–154, 158
gentrification 7, 9, 74, 76, 79, 131–148, 234

H
health inequality 8, 106–130
housing 9, 107–108, 132–148, 150, 164–194

L

lifestyle 7–8, 74, 76, 79, 164–194

M

multiple correspondence analysis 8, 17, 19, 25, 27, 46, 48, 69–82, 84, 91, 101, 208

N

networks 7, 9, 69–70, 73, 78, 85, 88, 138, 140–141, 197

P

principal component analysis 7–8, 13, 21, 24, 29, 73, 111, 117

R

rating scale 10, 213–226
religiosity 8, 83–105, 239
response style 213, 216, 224

S

social relations 8, 47, 69–82, 85
sociology 7–10, 74, 78, 132, 149–163, 164, 190–191, 195–212, 232, 241, 243
soviet union 9, 149–163
statistical education 9, 195, 199

V

visual priming 213, 224
vocational training 106, 127

Elina Fonsén | Raisa Ahtiainen
Marja Heikkinen | Lauri Heikonen
Petra Strehmel | Emanuel Tamir (eds.)

Early Childhood Education Leadership in Times of Crisis

International Studies During the COVID-19

2023 • 264 pp. • Paperback • 39,90 € (D) • 41,10 € (A)
ISBN 978-3-8474-2683-7 • also available as e-book in open access

The COVID-19 pandemic has dramatically affected all aspects of professional and private life worldwide, including the field of early childhood education and care (ECE). This volume sheds light on leadership in ECE: How did leaders experience the challenges they were facing and what coping strategies did they apply in order to deal with the changes in everyday life and practices in ECE centres? Authors from twelve countries present empirical findings gaining information on different crisis management mechanisms in ECE systems around the world.

www.shop.budrich.de

Yves Schemeil

The Making of the World

How International Organizations Shape Our Future

2023 • 406 pp. • Paperback • 44,90 € (D) • 46,20 € (A)
ISBN 978-3-8474-2146-7 • also available as e-book in open access

International Organizations (IOs) were designed to provide global public goods, among which security for all, trade for the richest, and development for the poorest. Their very existence is now a promise of success for the cooperative turn in international relations. Although the IO network was once created by established powers, rising states can hardly resist the massive production of norms that their governments can be reluctant to respect without being able to discard them. IOs are omnipresent, and exert great influence on the world as we know it. However, rulers and ruled are hardly aware of such compelling and snowballing processes. Yves Schemeil uses his in-depth knowledge of IOs to analyze their current impact on international relations, on world politics, and their potential of shaping the global future.

www.shop.budrich.de

Simone Seitz
Petra Auer
Rosa Bellacicco (eds.)

International Perspectives on Inclusive Education

In the Light of Educational Justice

2023. 268 pp. • Paperback • 59,90 € (D) • 61,60 € (A)
ISBN 978-3-8474-2698-1 • also available as e-book in open access

International developments and impulses call for the equitable and inclusive design of education systems. This book takes this up and focuses on the often blurred relationship between inclusive education and educational equity. By compiling current research results and theoretical contributions from several European countries on the topic, the authors create an overarching framework for discussion.

www.shop.budrich.de